To

MUM & DAD.

Happy 40th Anniversary,
with all our love

From

Carole, Norm, Bob & Scott

X .

NEW ZEALAND
NORTH TO SOUTH

NEW ZEALAND
NORTH TO SOUTH

TEXT GEOFF CHAPPLE
PHOTOGRAPHY JOHN COBB

VIKING PACIFIC

Title-page photograph: *Taieri River-mouth, Dunedin.*
Introduction: *Giant Gate Falls Milford Track, Fiordland.*
Contents: *Marakopa farmland.*

VIKING

Penguin Books (NZ) Ltd, 182-190 Wairau Road, Auckland 10, New Zealand
Penguin Books Ltd, 27 Wrights Lane, London W8 5TZ, England
Penguin USA, 375 Hudson Street, New York, NY 10014. United States
Penguin Books Australia Ltd, 487 Maroondah Highway, Ringwood, Australia 3134
Penguin Books Canada Ltd, 10 Alcorn Avenue, Toronto, Ontario, Canada M4V 3B2
Penguin Books Ltd, Registered Offices: Harmondsworth, Middlesex, England
First published in 1991
3 5 7 9 10 8 6 4

Designed by Suellen Allen
Printed in Hong Kong

ISBN 0670 841307

Additional photography by:
Kenneth Beatson; 26 (*above*), 126 (*below*).
Geoff Chapple; 27 (*above and below*).
Stephen Robinson; 70 (*above*).
Conrad Sims; 34 (*right*), 52 (*below*), 70 (*below*), 89 (*left*), 92 (*right*), 93 (*above*), 95 (*above*), 170 (*right*), 171 (*below*).

CONTENTS

Introduction		7
Key to Maps		9
1	NORTHLAND	11
2	AUCKLAND	23
3	WAIKATO	33
4	TARANAKI	41
5	COROMANDEL	49
6	BAY OF PLENTY	57
7	CENTRAL PLATEAU	67
8	EAST CAPE	77
9	HAWKE'S BAY	83
10	WELLINGTON • WAIRARAPA • MANAWATU	91
11	MARLBOROUGH • KAIKOURA	101
12	CANTERBURY	109
13	OTAGO	119
14	SOUTHLAND	129
15	FIORDLAND	137
16	MOUNT COOK AND THE LAKES	145
17	WEST COAST	157
18	NELSON	169

INTRODUCTION

The Unfurling Land

New Zealand began to emerge in its present shape just 30 million years ago. In geological time it is an infant, alone amidst encircling seas.

On such young and isolated islands, no dinosaur or mammal ever dominated. Evolution's preference was for birds, which filled the niches elsewhere taken up by the fast-developing mammal world: even down to grazing the plains. In a land free of major predators, flightless birds evolved, and the moa, eating sufficient grass each day to feed a bullock, became the largest bird ever to walk the earth.

New Zealand was the last significant landmass colonised by humanity. In Maori legend, the mythical hero Maui fished the land from the sea. The northern island was Te Ika a Maui — the fish of Maui — and the southern island, Te Waka a Maui — his canoe. In Maori oral tradition, the historical hero Kupe is said to have sailed south from Polynesia in command of two canoes to discover the islands he called Aotearoa, Land of the Long White Cloud. The event is embellished by myth, its dates uncertain, but archaeology confirms scattered camps in the South Island and around Taranaki dating from the seventh or eighth centuries AD. This first population — known as the moa-hunters or the Archaic Maori — was supplanted by Polynesian immigration of the fourteenth century, which began a Classical Maori tradition of strong tribes, extensive agriculture, large-scale fishing, fortified villages (pa) and warfare.

This was the culture that all the early European explorers brushed against — usually at their peril. The first, in 1642, was the Dutch commander Abel Tasman, who traced a single line of western coast, named by Dutch map-makers 'Nieuw Zeeland'. In 1769–70, in a more thorough and scientific British investigation, James Cook circumnavigated the islands and mapped them.

Whalers and sealers followed Cook, then missionaries arrived. European colonisation had begun before Britain annexed the country in 1840; the North Island by signature of Maori chiefs on the Treaty of Waitangi, and the South Island by right of discovery. The accommodation by Maori tribes to those they called Pakeha was at first adept, but as settler population began to expand, pressure on land increased and war broke out in the 1860s.

Land remains the country's wealth. When timber and gold were exhausted, wool seemed the country's only durable export. Then, in 1882, the introduction of refrigerated sea transport gave life to a meat and dairy industry. The forest was further cleared by burn-offs, and New Zealand became a grassland.

The new nation developed a strong egalitarian tradition. In the 1890s, it was the first country to give votes to women and a pension to the aged. In the 1930s, it further developed a welfare state to guarantee the nation's health, housing for the poor, and benefits for the needy. Although the welfare system and state ownership of key industries was partly disassembled in the 1980s, New Zealand's role as social laboratory to the world remained with radical anti-nuclear legislation and a strong conservation movement.

Triggered by the loss of Britain as a guaranteed market for its primary produce, the country has diversified over the last thirty years. Horticulture and viticulture now flourish alongside the traditional agriculture. Extensive pine forests now supply a large pulp, paper and timber industry. Light manufacturing has expanded alongside a few enclaves of heavy industry.

New Zealand is roughly equivalent in size to the British Isles, but has a population of just 3.3 million. Its 1,600-kilometre length falls entirely within the temperate latitudes. The climate is equable, without sharp seasonal variation in the north, and made mild even in the colder south by the surrounding seas. The North Island is slightly smaller than the South Island but is far more populous, supporting some 2.5 million people. The South makes up the deficit in the splendour of its unpeopled mountain landscapes, its resorts and ski-fields.

The two main islands have a distinct and extraordinary geography. The North's most spectacular scenery is the crater-and-hot-pool-studded volcanic belt running on a northeast-southwest axis across the middle of the island. The South's most scintillating landscape is the alpine chain that runs almost the whole of its length. Both are the product of a mighty tectonic force below. New Zealand is on the rim of the Pacific Plate. For millenia the country's mountains have steadily risen, its volcanoes have erupted, and its basement rock snapped in earthquake as the expanding plate planes downward into the earth's mantle.

New Zealand has adopted various emblems to show its character to the world. The kiwi, its most distinctive bird, is a popular symbol. So too is the koru. It is a green symbol of the tree-fern whose fronds now uncurl not in the ravaged bushland of the past, but in the immense protected regions of New Zealand's national parks. It is a blue symbol of the surrounding oceans and of the waves that unscroll continuously on New Zealand's long and pretty coastline, giving the country its distinctly maritime character. Most of all, it is the sign of a land still young, still unfurling.

KEY TO MAPS

North Island

South Island

Stewart Island

NORTHLAND

The landscape of the North Island's lean tail is hauntingly beautiful, and its history powerful. The first big Maori settlements were here, the first European towns, and the first signing of the treaty that gave birth to today's nation. The Muriwhenua, or 'Land's End' Maori tribes, sum it up in a saying: 'The flick of the tail guides the fish.'

Above: *Ninety Mile Beach.*
Left: *Striped marlin, skipjack tuna and shark at Tutukaka.*

WAITANGI

On 6 February 1840 the Treaty of Waitangi joined the destinies of Maori and Pakeha New Zealanders. It was first signed on the Treaty Ground outside the Treaty House (*right*). The house architecture is Georgian, the interior (*above*) Victorian.

The Maori developed Polynesia's most complex culture. Their carving, weaving and architecture express a rich symbolism. Tribal buildings invoke ancestral mana and protection. The Whare Runanga, or Maori meeting house at Waitangi, showcases such traditional skills (*above*). The pou-toko-manawa (*left*) supports both the ridge-pole and the ancestral line of Ngapuhi into an upper darkness.

New Zealand's most northerly reach is the Aupouri Peninsula. When the mists arise upon it, or storms sweep across, there is no better place to sense the land's fragility, the immensity of surrounding oceans, and the power of Maori tradition.

In that tradition, the western beaches are a pathway to spirits of the dead. At Te Rerenga-wairua, the last cliffs of Cape Reinga, the spirits plunge to the underworld, emerging in the fabled isles of Hawaiiki.

South from the cape, the road passes small townships and one of the largest exotic forests in New Zealand, the 20,000-hectare Aupouri Forest. At the base of the peninsula is the Far North's main town, Kaitaia. On the eastern coast is Doubtless Bay, where, over a thousand years ago, the Polynesian explorer Kupe is said to have first set foot on Aotearoa. True or not, the story fits the north's status as the cradle of New Zealand history. From the fourteenth to the eighteenth century, it supported New Zealand's largest Maori settlements, and in the nineteenth century, the first European towns.

Further down State Highway 10, past the citrus-growing town of Kerikeri, the beautiful Bay of Islands was host to those first Europeans. Barely 20 years after James Cook, whaling ships began to call, trading clothing, nails and muskets for food. The town of Kororareka grew – a bawdy hellhole that attracted a more saintly set: the missionaries. The Reverend Samuel Marsden, New Zealand's pioneering churchman, arrived at the Bay of Islands in 1814 to preach the country's first Christian sermon.

Final destination of the drive north is the Cape Reinga lighthouse (*above*). The lighthouse overlooks Te Rerenga-wairua, in Maori tradition, the jumping-off place of the dead. Cape Maria van Dieman (*below*) is part of the panorama. It was named by the seventeenth-century Dutch voyager Abel Tasman to honour a governor's wife in the Dutch East Indies.

A haphazard colonisation of New Zealand had begun. Lawlessness increased and missionary pressure forced a reluctant Britain to annex the country. Royal Navy Captain William Hobson was despatched to do the job, but to do so only with Maori consent. That was forthcoming in the famous Treaty of Waitangi, first signed by Maori chiefs on 6 February 1840, and subsequently taken around the country. It guaranteed to the Maori the lands, forests and fisheries they possessed, except for willing sale to the Crown. The chiefs were accorded the rights and privileges of British subjects, and in return ceded the country's sovereignty to the British Crown. The Treaty was New Zealand's founding document.

Kororareka's name was later changed to Russell. Today it is a pretty and historic town, a base for launch excursions around the Bay of Islands and for deep-sea fishing. Across the water is the Waitangi Treaty House and its historic grounds.

On the western coast the deep-set and sombrely beautiful Hokianga Harbour also supported early European settlements. Here, in 1838, the first Catholic Mass was celebrated by the French Bishop Jean Baptiste Pompallier. Small towns dot its shore, Rawene the biggest of them.

Great kauri forests once covered much of the north. The juvenile tree has no peer for standing straight and tall, and was prized in the days of sail for masts and spars. Nor is there a timber tree in the world to match the mature kauri's immense and branchless girth. The tree's desirability was its downfall. In the early 1800s, it was New Zealand's main export. South of the Hokianga, down Highway 12, remnants of the great forest are preserved at Waipoua Kauri Forest and the nearby Trounson Kauri Park. The tree provided another source of wealth: kauri gum. In the late eighteenth and early nineteenth centuries, it was dug from the earth and exported. The history of the kauri and the gum is displayed at the Otamatea Kauri and Pioneer Museum at Matakohe, further down Highway 12.

The small town of Houhora boasts the northernmost pub in New Zealand. Its wharves are host to the fishing boats (*above*) that bring in snapper, trevalli and kingfish to on-shore freezers. While advertising shellfish from the far end of the country, the Mangonui fish shop (*below*) also sells, seven days a week, the catch brought in fresh by the local fleet.

With timber and gum exhausted, the north turned to farming and fishing. The Northland Co-operative Dairy Company, with plant at Maungutoroto, Dargaville and Whangarei, is the second-largest dairy company in New Zealand, and one of the world's top milk-powder producers.

Whangarei (pop. 40,000) is the north's only city and serves the farming hinterland. On the south side of its harbour, the deepest non-dredged port in New Zealand serves the Marsden Point Oil Refinery, which processes 4.5 million tonnes of crude oil annually to supply almost all of the country's petroleum product needs.

At Waipu, south of Whangarei, Highway 1 passes a fine lion-topped memorial to the town's pioneers, Scottish Highlanders led by a Calvinist visionary through Nova Scotia to settle, finally, here.

The north is full of such history, of struggle to survive upon the land, but one wealth remains always: its beauty. No coastline in New Zealand matches the immense sandy sweeps of the western seaboard's Ninety Mile Beach or the indented succession of harbours and pohutukawa-hung bays on the east. Such beauty, coupled with a climate touted only a little inaccurately as 'winterless', makes the north a popular holiday destination.

And the tail still, occasionally, flicks. In 1988 a case brought by the Muriwhenua tribes before the Waitangi Tribunal forced a new and historic recognition that Maori retained some traditional ownership rights over New Zealand's commercial fishing grounds.

The history of Kerikeri township has a litany of pioneering firsts. They include the first plough to turn New Zealand soil, the first grass seed sown, the oldest wooden house in the country and the oldest stone building. The Stone Store (*below*) was built between 1832 and 1838 to store mission supplies. It is now open to the public as both store and museum.

The many sequestered bays and settlements of the Bay of Islands provide a wide variety of sights. In a subtropical pocket to the north lie Kerikeri's famous citrus orchards, (*right*). At Waitangi are the displays aboard Kelly Tarlton's Shipwreck Museum, (*above*) and across the water from there, historic Russell (*below*) reflects more peacefully now on a once-violent past.

Shadowed by a Moreton Bay fig, the Russell police station (*above*) dates from 1870, when it served as the port's customs house. On the waterfront, outriggers on the fishing boats (*centre*) mark Russell as a deep-sea fishing port, serving off-shore grounds that author Zane Grey once called 'the angler's El Dorado'. At the Captain Cook Memorial Museum (*below*) the seven-metre model of Cook's *Endeavour* summons early New Zealand history. The barque was the first European vessel to sail into these waters. Its famous captain named the Bay of Islands.

Whangarei Heads, northern arm of
Whangarei Harbour, offers protected
bays and beaches. An ancient
volcanic plug rises behind placid
McLeods Bay (*right*) on the
harbour's interior. Facing the open
sea, the wilder Ocean Beach (*below*)
is popular with surfers.

The North's past is evoked at the Otamatea Kauri and Pioneer Museum. Displays include the transport of giant kauri logs out of a rugged interior (*above*) and the pit-sawing of the timber in bush camps. Nearby Baylys Beach (*left*) attracts those who still live directly off the land's bounty.

New Zealand's grey giant, the kauri, (*right*) once colonnaded the North. The kauri forest's overarching beauty was compared by settlers to vast cathedrals, but the tree's immense branchless girth made it a valuable timber tree. For over eighty years the forests were milled and now only remnants remain. Waipoua Kauri Forest and Trounson Kauri Park are accessible reserves. The tree is easily identified by the hammer-beaten quality of its bark (*above*).

From the western seaboard the Hokianga Harbour reaches deep inland (*below*), its towns and settlements still carrying a pioneering air. At Rawene (*left*) is the Harp of Erin Hotel. At Opononi, a sculpture (*above*) commemorates Opo the dolphin. In the summer of 1955/56 Opo befriended curious bathers. For the second time in history a wild dolphin encouraged people to play. A law was passed to protect the dolphin, but like her historical counterpart in the Roman town of Hippo in AD100, Opo was killed.

AUCKLAND

In all things but politics and propriety, Auckland has an easy dominance as the biggest, the untidiest, the most pleasurable New Zealand metropolis. It is the nation's narcissus, the Queen City, where close to a third of the country's population lives, works and plays beneath mild skies, amidst a sea-struck landscape.

Above: *Auckland Harbour Bridge.*
Left: *Windsurfer at Okahu Bay.*

The business city stands tall from every ridge. Mirror glass of the Automobile Association Centre in Albert Street (*above*) reflects Queen Street high-rise. Office buildings (*right*) cluster above Grafton Gully. The view from Mount Victoria (*below*) on Auckland's North Shore, looking across the Waitemata Harbour to the city.

Auckland's history, like the land on which the city grew, is volatile. It sits on a natural minefield. Over a period of 60,000 years pocket-sized volcanoes have regularly erupted – all now extinct except perhaps for the last and largest of them, Rangitoto, the volcanic island at the mouth of the Waitemata Harbour.

From the fourteenth century, a succession of Maori tribes have occupied the isthmus. Its twin harbours, the Waitemata on the east and the Manukau on the west, together with the sheltered waters of the Hauraki Gulf, made it a prized territory. The volcanic cones provided sites for fighting pa, and fierce warfare gave the isthmus its ancient name – Tamaki makau rau, the land sought by a hundred lovers. Tribal battle continued as late as 1821, when musket-armed Ngapuhi tribesmen from the north slaughtered close to a thousand Ngati Paoa at the foot of Maungarei (Mount Wellington).

Within a few weeks of the first signing of the Treaty of Waitangi, New Zealand's Lieutenant-Governor, William Hobson, sailed south seeking a site for the new colony's capital city. He chose Tamaki makau rau, and on 18 September 1840, British officials bought 1,200 hectares of land from the occupying Ngati Whatua. The British flag was run up and the name of Auckland, an Admiralty lord and friend to Hobson, was carved onto the pole.

From the start, Auckland was different. It had no planned immigration, as did the other, more socially cohesive New Zealand cities. It had more Irish, more Maori, more grog shops, more South Seas riff-raff, more speculators and merchants. After it lost capital-city

Most of Auckland's volcanic cones were once terraced and fortified by bank and ditch defences into fighting pa. Maungakiekie, or One Tree Hill (*above*), was the most formidable of them, one of the world's largest prehistoric forts. *Below*: the Hauraki Gulf volcanoes Rangitoto, (background), a mere 600 years old, and Browns Island (foreground), some 20,000 years old, show contrasts of weathering and dimension.

The gannet colony at Muriwai (*above*) began on offshore Motutara Island, then spilled across to the mainland, where it can be directly overlooked. All of Auckland's west coast beaches display nature in the wild. At Piha (*right and below*) rollers from the Tasman Sea surge up iron-sand beaches, to provide rough-and-tumble swimming and spectacular scenery.

status to Wellington in 1865, Auckland became purely a city of business. The only 'father' it acknowledged was John Logan Campbell, who beached a canoe on its shores in 1840 to become a liquor merchant and later a philanthropist.

Always strongly influenced by the Maori, it now has 100,000 citizens of Maori blood. Its early missionary and trading connections with the Pacific have been amplified by waves of Pacific Islands immigration, and its population of Pacific Islanders is now 85,000. The two figures together give Auckland its often-quoted status as the largest Polynesian city in the world.

In physical size the Auckland metropolitan region ranks with the biggest cities in the world. Its 920,000 population, by no means a world ranking, has simply sprawled. An organisational reform of 1989 gathered the region into four separate civic entities – on the isthmus, Auckland City; to the north, North Shore City; Waitakere City to the west; and Manukau City to the south.

Manukau City, reached by the Southern Motorway, is the fastest-growing urban area in New Zealand. Much of the region's industry is sited there, including its main sewage treatment plant, electricity substations and the Auckland International Airport.

Waitakere City, reached on the Northwestern Motorway, is a fruit-growing and wine-producing region, gateway also to the Waitakere Centennial Memorial Park with dozens of bush walks. Beyond the Waitakere Ranges lie the popular west coast surf beaches.

North Shore City, accessed on Highway 1 across the Auckland Harbour Bridge or by ferry across the harbour, is a seaside metropolis. Its long east coast features a succession of calm, island-protected beaches.

Performers dressed in coprosma leaves (*above*) at the Auckland Secondary Schools Maori and Pacific Islands Festival. Wearing traditional garb, another group prepares for performance (*below*). Polynesians make up one-fifth of Auckland's population. More recent immigration has swelled the total to give Auckland its claim as the largest Polynesian city in the world.

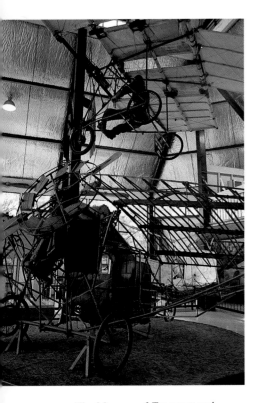

The Museum of Transport and Technology includes poignant relics (*above*) of Richard Pearse's genius. Some credit the Temuka farmer with beating Orville Wright into the air in 1903. The neo-classical Auckland War Memorial Museum (*below*) houses New Zealand's finest Maori collection.

Auckland City occupies the isthmus. It is centred around Queen Street, a high-rise canyon of business houses and retail shopping. Its layout is best studied from the public lookout atop the State Bank tower in Queen Street, or from Maungawhau (Mount Eden), a volcanic cone that rises behind the central business district and, at 196 metres, the highest hill on the isthmus.

From the downtown Ferry Buildings regular launch trips leave for islands of the Hauraki Gulf, one of the longest protected waterways in the world and a paradise for sailing craft. Any ferry trip onto the gulf reveals Auckland's spinnakered soul. The city offers no finer sight than the thousands of yachts, power craft, wind-skis and kayaks of maritime Auckland cleaving the harbour and gulf on a summer weekend, or packing it almost to a standstill on regatta days.

Most of the eleven gulf islands are part of the Hauraki Gulf Maritime Park. Two of them, Waiheke and the more far-flung Great Barrier, are outside the park and have tourist accommodation. One, Pakatoa, is a privately owned resort. Steep-sided Little Barrier is a protected wildlife sanctuary.

North of the city on the main highway are the Orewa Hot Springs, the lovely Wenderholm Regional Park, Puhoi (a Bohemian settlement dating to 1863), and the Goat Island Marine Reserve near Leigh. North up Highway 16, near Helensville, steam the Parakai Hot Springs.

South of the city, a turn-off east from State Highway 1 at Papakura leads to the shallow Firth of Thames, a coastal drive suitable for bird-watchers, and hot springs at Miranda.

A turn-off west from Highway 1 at the top of the Bombay Hills leads on past the market-gardening town of Pukekohe to reach the Waikato Heads, where the Waikato River discharges to the sea. From there each year 1.5 million tonnes of iron-sand concentrate are pumped to the distant Glenbrook Steel Mill, which produces 750,000 tonnes of steel annually. A road leads back up the coast within sight of the mill to end at the Manukau Heads and a view of powerfully surging tides across the bar at the Manukau Harbour entrance.

Each New Year's Day the city's richest horse race, the Auckland Cup, is run at Ellerslie Racecourse (*left*). Other holiday pleasures include fishing at Great Barrier Island in the Hauraki Gulf (*above*). The island is New Zealand's fourth largest, and is sparsely populated, offering scope for tramping along a deserted coastline (*below*) or into the interior.

HISTORIC BUILDINGS

The Puhoi pub (*above*) is the social centre of a Bohemian settlement north of Auckland, founded in 1863 by immigrants from the Czech-German border region. Highwic House (*above right*) is a fine example of the Gothic Revival style popular in early Auckland. Built in 1862, this historic home is open to the public. Alberton (*right*) dates from 1862. It was once the elegant home of a gentleman farmer. Mansion House (*below*) on Kawau Island, has the most interesting history. Its owner rode about his island in a coach pulled by zebras. He was no local eccentric, but New Zealand's most powerful nineteenth-century statesman, Sir George Grey, who at times administered the nation from his island drawing room.

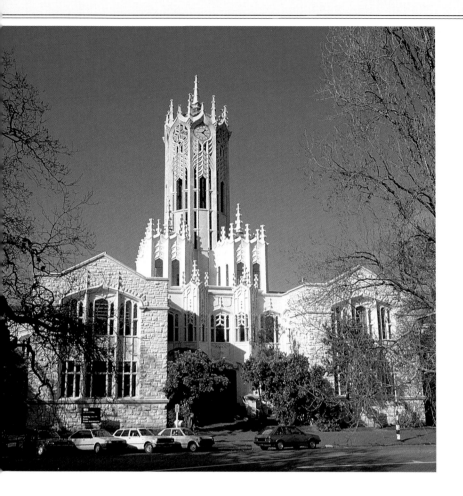

The Gothic spires of the Auckland University clock tower (*left*), commonly called 'the wedding cake', have been a city landmark since 1926. Parnell buildings (*above*) are amongst Auckland's oldest, for Parnell Rise, immediately east of the city centre, was the first suburb settled. It features a village of restored Victorian buildings, and sufficient Italian restaurants to earn it the local title 'Parnelli'. The Auckland City Art Gallery (*below*) is an important keeper of the country's painting tradition. Half its permanent displays are New Zealand art. The architectural style shows a sylistic mix, once described as 'Victorian revivalist interpretations of sixteenth-century French mannerist architecture'.

WAIKATO

The grass grows no greener than in the Waikato Basin. Nor does any New Zealand waterway match the river that runs in slow motion through it: the mighty Waikato. In this limousine of lands the cow cocky is king, but southwest the land tumbles toward beef cattle country, the famous Waitomo Caves and a wild west coast.

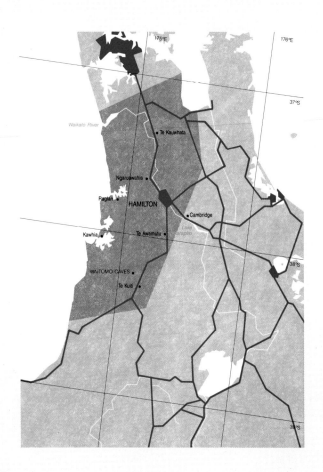

Above: *The Waikato River.*
Left: *The Hamilton river bridge.*

A Maori ceremonial canoe on the Waikato River (*right*) during celebrations at the Turangawaewae Marae, Ngaruawahia, residence of the Maori monarch. Power boat racing (*below*) on Lake Karapiro. The smooth water and 24-kilometre reach of the artificial lake makes it a favoured venue for all flat-water sports, and the headquarters of New Zealand rowing.

From trickling, snow-fed headwaters at the country's volcanic heart to its broad discharge on the west coast, the Waikato River flows 450 kilometres to the sea. It is New Zealand's longest, an ancient progenitor of the Hauraki Plains, and later, its course changed by an cataclysmic upheaval, the river laid that same alluvial wand upon the Waikato Basin.

On such fertile lands, and beside a river that opened to their canoes a trading path deep into the heartland of the North Island, the Waikato Maori prospered. Such strength and self-sufficiency kept the region secure as a Maori stronghold, but the European glance was covetous.

In the 1850s, on a trade route that travelled downriver, briefly overland to Waiuku and across the Manukau Harbour, Waikato Maori supplied Auckland with food. Maori farms and flour mills dotted the Waikato's banks. By then the European population was outstripping the Maori one, pressure for land was intense, and Maori resistance to European dominance culminated in a loose federation of Waikato, Taupo, Hawke's Bay and southern Taranaki tribes, under a Maori King.

War followed. Gunboats sailed up the Waikato in 1863 to support a bloody assault on the Maori fortress at Rangiriri, and took the Maori King's pa at Ngaruawahia without a fight. Another pitched battle was fought against Rewi Maniapoto at Orakau, and the Waikato campaign ended with large-scale land confiscation and the establishment of armed settlements on the river banks.

The Waikato River flows past the Huntly power station (*below*). Using cooling water from the river, and local coal, the station generates 1,000 megawatts of electricity, making it the country's largest power station.

The Karapiro power station (*above*) was commissioned in 1947, the largest of ten hydro-electric dams on the Waikato. Agriculture Field Day at Mystery Creek near Hamilton (*below*).

Down Highway 1, past coalfields that fire the thermal power stations at Meremere and Huntly, the city of Hamilton (pop. 104,000) began as an armed settlement. Part of its preserved tradition is the gunboat *Rangiriri*. Built especially for the Waikato campaign, it now sits meekly at the riverside in Parana Park. Hamilton is New Zealand's largest inland city. It has never lost its tough farming streak, and rugby football is here close to religion.

As centre of a wealthy farming region, civic wealth has accumulated too, but only over the past 30-odd years has it spilled into aesthetic embellishment and culture. The city's lovely fountains, Waikato University, the Art Museum, and the Founders' Memorial Theatre were all established within that time.

Hamilton's spirit and that of the entire Waikato is perhaps best assessed during the Agricultural Field Days at Mystery Creek, when animals and farm machinery are on display, and which reach the pitch of a celebration. The nearby Ruakura Agricultural Research Centre, at the forefront of investigation into animal genetics, health and production, is a part of that same tradition.

South of the city is Cambridge, an oak-shaded stud-farm centre, also once an armed enclave.

Runholders on large estates drained the Waikato swamplands, then small-holders further groomed the land into the sleek fields of today. The Waikato Basin supports cropping, including vineyards at Te Kauwhata, deer and goat farms, but the basis of its wealth is 600,000 dairy cows, whose butter, under the now famous Anchor brand of the New Zealand Co-operative Dairy Company, was first exported in 1888.

In the 1940s, work began to harness the Waikato River for hydro-electric power. Karapiro, on Highway 1 south of Cambridge, is the largest of eight dams on the river. Its artificial lake is 24 kilometres long.

The entrance to Raglan Harbour (*left*) is a beauty spot of the Waikato's west coast, and offers some of the coast's smoothest surf. The coastline near Kawhia, looking across Ocean Beach (*above left*), encloses an idyllic harbour. At Kawhia (*above*) two residents prepare to go mullet fishing on the best but least developed of the North Island's west coast harbours.

The King Country's most scenic
road leaves Highway 30 just before
Te Kuiti and winds 50 kilometres to
the coast and the tiny fishing
settlement of Marokopa (*right*). The
road passes the beautiful 36-metre
Marokopa Falls (*above*) and the
Waitomo Caves. Limestone country
all about is riddled with caves, but a
group of three have become a great
tourist attraction, serviced by a
small settlement and a grand old
hotel. There is the Waitomo Cave
itself, famous for its glow-worm
grotto; Ruakuri Cave, which is the
largest; and Aranui Cave (*below*),
which was the last discovered, in
1910, and which many consider the
most magnificent.

From Hamilton, Highway 23 leads out to a beautiful coast at Raglan Harbour. Highway 3 leads southwest through the farming town of Te Awamutu to a land formidably crumpled and toothed with limestone bluffs. At the end of the Waikato war the Maori King took refuge here. The area was never conquered and became known as the King Country, but King Tawhiao emerged in 1881 to make a formal peace and reclaim Ngaruawahia as the headquarters of the Waikato tribes.

In the 1880s, an unleashed Pakeha enterprise swept the hills. The King Country forests were milled, the Main Trunk railway line pushed onward, and the region finally settled on a sheep, cattle and pine-forest economy that supports the region's towns at a modest level. Pureora Forest Park preserves virgin bush for hunting and tramping.

Westward from Otorohanga, on Highway 31, is Kawhia Harbour. According to tradition, Kawhia is the last resting place of the *Tainui* canoe, which brought the Ngati Maniapoto and Waikato tribes to Aotearoa. Although this is the finest natural harbour on the North Island's west coast, Kawhia's waters are fished and boated by only a handful of residents, their semi-secret idyll further enhanced by the thermal waters of Te Puia Springs, which seep through the sands of Ocean Beach.

Just before Te Kuiti, a side road turns right from Highway 3, heading through limestone country to the Waitomo Caves, one of New Zealand's natural wonders. Hundreds of water-hollowed caverns exist, many as yet unexplored or merely glimpsed in adventure trips like the newly popular blackwater rafting. Three main caverns have been developed for the tourist trade. The Waitomo Cave is world-famous for its glow-worm grotto; the Ruakuri and Aranui Caves are known for limestone accretions weirdly, wonderfully, naturally, gothic.

Cattle-grazed grasslands near the coast in south Waikato (*left*). Since 1900 the north Waikato town of Te Kauwhata has been a wine-growing district, its government viticultural research station now joined by private vintners like Cook's Winery (*top*). The rose is the symbol of the south Waikato town Te Awamutu. The town's cherished rose garden (*centre*) features over 2,000 bushes, and some eighty varieties. Autumnal colours at Te Kuiti (*above*), a farming town in the King Country.

TARANAKI

The solitary cone of Taranaki the mountain is touchstone to Taranaki the province. The mountain laid down the rich soils of the region's pasturelands, and waters them still with an abundant rainfall pulled from the sky. Mount Taranaki endows New Zealand's smallest and richest province with a striking beauty and a strong identity.

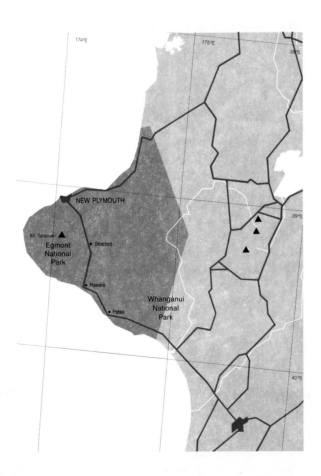

Above: *Mount Taranaki.*
Left: *Taranaki farmland.*

Now no more than two-million-year-old stumps, the Sugar Loaf Islands off the coast of New Plymouth are the oldest remnants of Taranaki Province's volcanic past. The Kaitake and the Pouakai Ranges, too, are both heavily weathered volcanic massifs, once perhaps as imposing as the single volcano whose soaring 2,518-metre symmetry now dominates the province, Taranaki the mountain.

The volcano erupted first some 120,000 years ago, building a cone hundreds of metres higher than today's summit. Subsequent mudflows from the giant peak have determined much of Taranaki Province's surrounding landscape, a vast apron of volcanic soil, the tephra from past eruptions adding fertility.

The mountain's volatile character is poetically preserved in Maori myth. Taranaki is said to have stood once in proud company with the mountainous warriors of the Central Plateau, but after a quarrel with Tongariro, Ngauruhoe and Ruapehu over the lovely female mountain Pihanga, Taranaki was driven into exile. Today the mountain sleeps, but is not extinct. Pumice from its eruptions can still be found lodged in old trees. The mountain's last increase in height occurred around 350 years ago, when a lava plug popped up from the central crater, perhaps explaining a further Maori myth: that Taranaki swells visibly with pride after famous tribal victories.

Cattle near Stratford (*left*), and Mount Taranaki (*below*), looking south across the Pouakai Range. Both the mountain and the range are volcanic. The extinct Pouakai was once a volcanic mountain perhaps 2,000 metres high, and last active some 200,000 years ago. Mount Taranaki is much younger. Ash showers occur at century-long intervals, and the potential for a major eruption remains. The mountain is merely dormant.

The early history of the Taranaki area is obscure, but voyagers from Polynesia probably arrived around AD 700 developing a moa-hunting culture. The later oral tradition names three canoes, *Tokomaru, Aotea,* and *Kurahaupo,* arriving around AD 1300, as the ancestral waka of today's tribes.

The Dutch explorer Abel Tasman sighted the cape in 1642, but it was James Cook, standing out from the coast in 1770, who was the first European to see and describe, towering amidst storm gloom and lightning flash, a mountain of 'prodigious height . . . its Top covered with everlasting snow'. He named the peak Mount Egmont, which still adheres as an alternative to the Maori name.

European settlers first arrived aboard the *William Bryan* in 1841. By the 1850s, Taranaki was the centre of a dangerous tension. Suspect land purchases set Maori against Pakeha and Maori against Maori as the imported notion of individual ownership fractured traditional tribal ownership and loyalties. In 1859, Governor Gore Browne denied the right of an important chief to veto a sale agreed to by a lower-ranking Maori. This dispute, over open coastal land at Waitara, sparked the land wars of the 1860s and subsequent large-scale confiscation of Maori land. Tension between the races continued after the last Taranaki conflicts of 1869. Passive resistance against settler incursion on disputed lands was developed as a peaceful weapon by the Maori prophet Te Whiti O Rongomai. His farming community at Parihaka was dispersed by Armed Constabulary in 1881, but the marae still exists at the western foot of the mountain.

Taranaki has developed into the richest dairying province in New Zealand, afloat on milk,

The observatory at Cape Egmont (*above*). The faraway peak of Mount Taranaki completes an Elysian idyll at New Plymouth's Pukekura Park. The romantic 'Poet's Bridge' across Pukekura Lake commemorates more than poesy and was built literally by chance. In 1883 a park trustee bet on a horse race, boasting he would give any winnings to park development. The horse came in. Its name was The Poet.

but today also on oil. The first well was sunk in 1865, though Taranaki's status as the energy province of New Zealand was not clear until the 1950s, when the Kapuni gas field was tapped, followed by discovery of the large offshore Maui gas field, and the commissioning of a methanol plant, and the gas-to-gasoline synthetic fuel plant at Motunui in the 1980s. Over 30 per cent of New Zealand's petroleum products now issues from the gas and oil fields here, along with natural gas, which is piped throughout the North Island.

Prosperous New Plymouth (pop. 48,000) is the region's only city, but the rich countryside is more extensively settled than farming country elsewhere in New Zealand and supports many market towns. New Plymouth's Pukekura Park and the adjoining Brooklands Park form one of the country's most beautiful reserves, combining highly wrought fountain displays with boating pleasures on the lakes and a surrounding rainforest magic.

Pukeiti Park, 29 kilometres from New Plymouth, is another beauty spot. Gardens of azaleas and rhododendrons, primulas and alpine plants are on display amidst groves of exotic and native trees. The park abuts the greatest of Taranaki's reserves, the giant radius of the Egmont National Park. The mountain is its centre, and the wilderness of alpine flora, swamp tussock and rainforest below is threaded with over 300 kilometres of track, with huts spaced three to seven hours apart for round-the-mountain trampers, making it one of the most accessible of the New Zealand national parks.

Taranaki is the most climbed of the country's snow-capped peaks, though it can be dangerous and advice should be sought from the North Egmont Visitor Centre or one of the lodges before attempting the five-hour ascent to the summit.

Ngamotu golf course at New Plymouth (*top*). Farmland (*above*) and market gardens (*below*) in south Taranaki.

Taranaki is the most densely settled of New Zealand's rural regions. Its dairying land is cross-hatched by small farms and dotted with small towns. Yet inland from the smooth volcanic plains surrounding the mountain, the geography changes. The land climbs toward the Central Plateau, in tumbling backblocks country behind Mount Messenger (*above*). These razorback ridges, looking northeast from the Mangawhero Valley (*right*), are sheep country. Back on the salty coastal flats near Opunake, a dairy farmer (*below*) on a four-wheel-drive motorbike, feeding out hay.

The styles and statements of Taranaki's buildings vary widely. The 50-metre water tower at Hawera (*above*) is the town's most prominent landmark, and features in the novels of Ronald Hugh Morrieson, a New Zealand writer. The striking symbolisms of the Ratana church (*above left*) in Ratana township mark the birthplace of a messianic Maori religion. The date on the arch commemorates the appearance of an angel to Tahupotiki Wiremu Ratana, telling him to unite under Jehovah Maori tribes weakened by the impact of colonial rule and disease. The townspeople still preserve relics of Ratana, and celebrate his 25 January birthday with a festival. Against such a folk-driven style, the neoclassical portico and clock tower of Patea's library (*centre left*) stand in sharp contrast, while a Maori church at Tutahi (*left*) displays New Zealand's traditional weatherboard and corrugated-iron construction.

COROMANDEL

The indented coastline of Coromandel Peninsula yields a pocket paradise at every bend of the road. Red-flowered pohutukawa overhang pebble-grey coves and blue bays. Behind, green land rises steeply to black peaks. The peninsula's past is more colourful yet, its valleys once raucous with the boom towns of gold and silver mining.

Above: *Otama Beach.*
Left: *Mercury Bay.*

The Coromandel's rugged spine is volcanic. Twenty million years ago the volcanoes had height and fury sufficient to spread an iron-rich breccia as far as Auckland, where it is still visible as a rust-brown strata in the cliffs. Even today the weathered and sometimes grotesque peaks of the range top 800 metres. Within their flanks are heat-juiced minerals from the earth's core, including big gold- and silver-bearing quartz 'blows'.

According to tradition, Kupe the Polynesian explorer was first to sight the peninsula. Later, Maori from the migratory canoe *Arawa* stayed, and the captain, Tamatekapua, is said to be buried near the summit of the tallest peak, Moehau (892 metres).

The Marutuahu tribe from the Waikato later conquered most of the peninsula. It was such Maori who watched 'goblins' – white men uncannily rowing with their backs to the destination – come ashore from a sailing vessel anchored just northeast of Whitianga. The date was 5 November 1769; James Cook had arrived to observe the transit of the planet Mercury. In that same fateful November, Cook explored and named the Firth of Thames on the inner side of the peninsula. He noted 'immense woods of as stout lofty timber as is to be found in any part of the world'.

The felling and export of these kauri forests became, from the end of the eighteenth century, the peninsula's first European industry. In 1820, a British Admiralty store ship, despatched

The surf at Ocean Beach (*left*) makes the nearby town of Whangamata a popular holiday resort. Thames (*below*) is the biggest Coromandel town. Its gold-rush origins are still apparent in stalwart Victorian architecture, and it stands today as a gateway town to the coastal pleasures beyond. The Thames Union Parish Church (*above*), built in 1898, displays the distinctive square Norman tower of the Early English style.

Coromandel's gold-rush history is preserved in mining museums at the main towns. At Waihi (*above*) the collection centres around the huge Martha Mine, recently reopened. At Coromandel (*right*) the museum includes photographs of the gold workings, mining relics, and – of great interest to rock hounds – a display of the peninsula's semi-precious stones. Coromandel township itself is a quiet settlement. Victorian homes suggest its former wealth, but more common are the distinctive baches (*below*) that give both town and peninsula a special character.

to load spars for the navy, sailed into the Waiau Harbour midway down the peninsula's western side and left behind its name – *Coromandel* – on what soon became a bawdy little timber and boat-building settlement.

In 1852, an American, Charles Ring, found auriferous quartz and gold dust in the Kapanga Stream at Coromandel, and 3,000 diggers crowded in to New Zealand's first, but relatively minor, goldfield.

The first really large strike occurred at Thames in 1867, when alluvial gold was quickly traced to mother lodes of quartz. Mining companies floated by eager Auckland capitalists took over the field from the first itinerant panners, and for years the earth shook to stamper batteries crushing ore from over seventy mines.

Other fields followed, all of them finally outshone by a strike at Waihi. There the churning poppet-heads of the Martha Mine, New Zealand's richest, extracted gold worth $4,500 million. At Waihi in 1912, a bitter strike culminated in the killing of a strike leader, the bloodiest of the early labour struggles. The mine closed in the 1950s.

The peninsula's enduring treasures are more subtle; the beauty of the coastline and the rugged wilderness of the interior. Coromandel Forest Park has a headquarters just beyond Thames, but extends to the tip of Cape Colville, with tracks that wind through kauri groves, lead on to old logging and mining sites, or ascend to craggy peaks. The peninsula is also known as the best fossicking ground in the North Island for semi-precious stones. Thermal processes have filled host stones with mineral dyes, and on beaches and in rivers shine the reds of carnelian, the greens of jaspar, the variegated yellows of agate, and even, at Tairua and Hikuai, the occasional irridescence of opals.

Though still located as occasional colour in streams, gold lingers mostly as a ghostly history in the mining museums at Thames, Coromandel and Waihi. The peninsula's towns

The road winds past the rocky shores of Colville Bay (*below*) around the base of Mount Moehau, Coromandel's highest peak, to reach Cape Colville, at the peninsula's tip.

On the peninsula's eastern seaboard is Whangapoua (*above*), here populated by a doughty band of oystercatchers. At Whitianga a local fishing fleet (*right*) takes advantage of the best natural harbour and fish-processing facilities on the eastern coast. The popular resort, lies in beautiful Mercury Bay (*below*). James Cook dropped anchor here in November 1769 and formally hoisted the British flag to claim New Zealand for the King of England. He and the scientists who had travelled with him aboard the *Endeavour* also observed the transit of Mercury, and so named the bay.

have had to find a life beyond the old timber and gold extraction, or perish. Thames (pop. 6,800) still thrives as a market town to the dairy-rich Hauraki Plains, and as entranceway to the peninsula. North on the Highway 25 loop road, Coromandel has shrunk to a population of around 1,000, mainly fishing, craft or retired people, but restoration of its colonial architecture is exemplary.

On the eastern coast, Whitianga, once a deep-water port for timber exports, is now a popular resort and a base for deep-sea fishing. Tairua and the exclusive subdivision of Pauanui are holiday havens on opposite sides of the Tairua Harbour. Whangamata, further down the highway, is another holiday mecca, known particularly for its four-kilometre surf beach.

At Waihi, the farming hinterland and initiatives in specialist electronics helped survival beyond gold. Then in 1988, the Martha Mine reopened. Using open-cast methods, it is again New Zealand's biggest goldmine, producing $40 million of gold and silver doré bullion annually. Other old mine sites are re-opening, and prospecting rigs, strenuously opposed by many locals, now dot the peninsula. Gold retains its potential to lift emotions to the boil, but with a late twentieth-century spin. The emotion is conservationist, concerned to preserve against any industrial encroachment the most beautiful finger of land in the North Island.

Sheltered by Mahurangi Island, Hahei Beach (*left*) is one of Coromandel's loveliest fishing, diving and sand-kicking bays. At Hot Water Beach (*top*) thermal water seeps underfoot, and between low and mid-tide, visitors can scoop their own hot pools in the sand. Resort development has taken place around Tairua Harbour (*above*). Tairua itself remains a traditional town, which burgeons to resort status in summer with ocean-surf or calm-harbour swimming and skin-diving attractions. On the distant southern arm of the harbour is Pauanui. Planned as a 'getaway' resort for the wealthy, it has its own airfield, golf course, security systems and a building code to maintain the tone of its holiday homes.

55

BAY OF PLENTY

James Cook, who named the Bay of Plenty in 1769, got it right. He was impressed by the number of Maori plantations onshore. Today, crops from fruit to flowers all thrive. New Zealand's volcanic belt runs across the region, breaking surface at spectacular Rotorua and again at the White Island volcano 50 kilometres out from the coast.

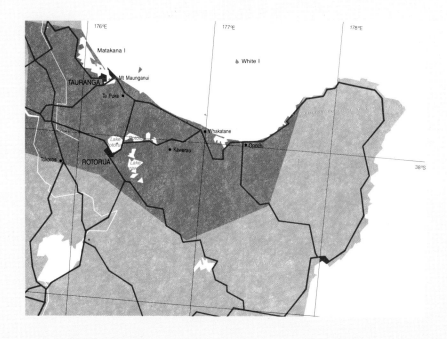

Above: *Ocean Beach, Mount Maunganui.*
Left: *White Island.*

FORESTRY

The Tasman pulp and paper mill at Kawerau (*above*) processes two million tonnes of wood every year, producing 350,000 tonnes of newsprint and 200,000 tonnes of market kraft pulp, most of it destined for export. The New Zealand Forest Products mill at Kinleith (*right*) is of comparable size and production. The two mills together produce the major part of New Zealand's pulp and paper exports. Pine forest at Atiamuri (*below*) is part of an immense hectareage of forest in the Bay of Plenty hinterland. Land once laid waste by Taupo's ancient eruptions now supports New Zealand's largest exotic forests, Kaiangaroa and Kinleith.

The Bay of Plenty hinterland, stretching into the central North Island beyond Rotorua, supports New Zealand's two biggest pine plantations: the Kaiangaroa State Forest (150,000 hectares) and Kinleith Forest (145,000 hectares). Each year, millions of tonnes of logs and timber from the hinterland towns of Tokoroa and Murupara, plus a million tonnes of pulp and paper from the mills at Kinleith and Kawerau, are trucked and railed to the distant port of Tauranga, making it the largest exporter in New Zealand.

The character of Tauranga city (pop. 63,000) is, however, more attuned to its surrounding orchards than the big pine-based industries of the interior. The city began as a missionary settlement, and in 1864 saw fierce fighting between Ngai Te Rangi Maori and British troops.

For the British, the battle of Gate Pa was one of the most disastrous of the land wars. The site is marked still by an Anglican church in Cameron Road, where the story is preserved of a Maori woman, Heni Te Kirikaramu, who took water to wounded British soldiers. Closer to the city centre early history is preserved in a cliff-side complex that displays the guns and trenches of the Monmouth Redoubt, the Tauranga Mission House and the Otemaha Pa military cemetery. Around that old centre the city is packed neatly and pleasantly, enjoying a mild climate and a splendid harbour.

The 232-metre cone of Mount Maunganui marks the harbour's southern entrance. It dominates a long sandy peninsula and the township of Mount Maunganui, the Bay of Plenty's most popular resort. Each summer thousands of holiday-makers invade the peninsula, percolating through the hot salt-water pool at the base of the mountain or plunging into the amiable surf of Ocean Beach. From the town, scenic flights depart for White Island, and launches make trips to the game-fish and skin-diving waters around Mayor and Motiti Islands.

The Tauranga Historic Village (*above*) is a pocket of Victorian nostalgia amidst the twentieth-century bustle of the city beyond. The shops are nineteenth-century and display the wares of the period. The streets are cobblestoned and echo to the clang of the blacksmith's anvil and the hammer of the wheelwright. Rides on an 1877 steam train are a popular 'live day' attraction. Port Tauranga (*left*) owes its status as New Zealand's biggest export port to the log, timber, and pulp and paper tonnages generated in the Bay of Plenty hinterland. Its tugs are named after those same hinterland towns.

The kiwifruit harvest at Te Puke (*above*) is the high point of the town's calendar, and celebrated by festivities. Kiwifruit is the Bay of Plenty's main horticultural crop and an important export. The vine was brought to New Zealand from China in 1906, but its potential was not realised until decades later. Hybrid strains were developed, yielding a larger fruit. A new name 'kiwifruit' replaced the older 'Chinese gooseberry' and, thus equipped, the furry-skinned export successfully stormed the luxury fruit markets of the world. At Mount Maunganui fishing trips (*above right*) are part of an extensive range of tourist services. 'The Mount', as it is commonly called, is one of New Zealand's top beach resorts. It incorporates the 232-metre cone at the entrance to Tauranga Harbour, and the long sandy peninsula (*centre right*), populated by crowds of 30,000 and more in summer, which connects the cone to the mainland. Matakana Island (*right*) forms a long and low protective barrier at the mouth of Tauranga Harbour. It is Maori-owned and planted out in commercial pine forest, with one small settlement on its western tip.

The furry-skinned, piquantly green-fleshed kiwifruit always grew well in the equable climate and rich volcanic soils of the Bay. Recently the world appetite for the fruit became voracious, and at Te Puke, commonly known as the Kiwifruit Captial, and on a 15-kilometre coastal strip stretching southeast as far as Opotiki, the land is now cross-hatched with shelterbelts of this horticultural king-maker. Nationally it is the most important fruit export, bringing in export receipts of over $500 million a year, and the Bay of Plenty accounts for 62 per cent of the crop.

Whakatane, the next major town on coastal Highway 2, has a mill producing cardboard from recycled paper, but the town is better known as a deep-sea fishing base and for the splendours of nearby Ohope surf beach. Southeast from there the road leads through Opotiki, gateway town to the East Cape.

Inland, on Highway 30, is Rotorua. New Zealand's most popular tourist town first hosted a steady flow of visitors in the 1870s after fighting against the Hauhau Maori and the rebel Te Kooti ended. Visitors journeyed out to Te Wairoa village and canoed across Lake Tarawera to see 'the eighth wonder of the world'. On the southwestern side of Mount Tarawera, delicately descending, were silica basins, brimming with water – the famous Pink and White Terraces.

Surges in Lake Tarawera, the appearance of a 'phantom canoe' and the warnings of an old tohunga presaged the biggest eruption in New Zealand's recorded history. On the night of 10 June 1886, the mountain was rent by eruptions. Scalding waves of mud swept away two Maori settlements, and Te Wairoa was buried under rock and ash-fall. Over 150 people died, and the Pink and White Terraces were destroyed.

A fisherman displays his catch on the beach at Opotiki (*above*). Whakatane (*below*) is, after Tauranga, the largest population centre in the Bay of Plenty. It has an industrial base, is the market town for surrounding sheep and dairy towns, and caters for holiday-makers and tourists with excellent swimming, big-game fishing and scenic flights.

Rotorua's Blue Baths were first opened in 1885, but fell into disrepair, and this 1930s building (*right*) designed by a Public Works architect and built by the Tourist Department, replaced it. An enthusiastic government had also funded the Rotorua Bath-house (*below*). It opened in 1908, but like all Rotorua buildings enclosing corrosive hydrogen sulphide gases from hot pools, the Tudor-style edifice became a maintenance nightmare. By the 1960s most of the pools were closed, and the building now functions mostly as an art gallery and museum.

Rotorua's fame then settled on its hot springs. The Rotorua Bath-house opened in 1908, an expensive Tudor-style building funded by a government that believed the area could become the greatest health spa of the Southern Hemisphere. Confidence in the curative power of its waters had waned by the 1940s when the Government Sanatorium closed and the Bath-house was a corroded victim of the hydrogen sulphide gas that rots metal and concrete alike throughout Rotorua. By the 1970s, the main hot springs were in private hands. Polynesian Pools renovated the main springs for its big recreational pools, but the complex includes 26 private pools and, a last vestige of the old faith, the Priest Bath, said to be good for rheumatics. The old Rotorua Bath-house still exists amidst the beautifully manicured Government Gardens, renovated now and housing an art gallery and one restored bathing pool.

Rotorua is today a city (pop. 54,000) and a popular tourist destination. Within easy reach of the city are four cold-water springs teeming with trophy-sized trout, eleven fern-fringed lakes, and, at the Agrodome, an instructive and entertaining display of New Zealand's sheep industry.

Most compelling, though, are the thermal regions. Whakarewarewa, on the southern margin of the town, is the most popular. New Zealand's biggest geyser, Pohutu, plays regularly here amidst hot springs by the hundred, mud pools and hissing fissures. At the adjacent Arts and Crafts Institute, Maori carvers and weavers practise the old skills and sell their work.

North of the city is Ohinemutu village, a Maori settlement with an historic church and graveyard, dotted throughout with hot pools and vents, many in domestic use. Tikitere, also known as Hell's Gate, is 18 kilometres east of the city. Its mud and boiling pools are furiously active, and visitors can splash beneath a warm waterfall. The Waimangu Valley, 20 kilometres south of Rotorua, is the most primal of the thermal regions. Its cliffs, rent by the eruption of Tarawera, steam with menace. A four-hectare boiling lake, created by a further eruption in 1917, is awesome. Accessible from here is the village of Te Wairoa, buried by ash in the Tarawera eruption and now partially excavated.

Waiotapu is 32 kilometres south of Rotorua and features an enchanting sheet of shallow water, with small geysers fountaining. The area is encrusted with silica formations called, for their wonderful range of colour, the Artist's Palette. Further south again, some 68 kilometres from Rotorua, is Orakei Korako, whose striking beauties include a pool lying deep inside a cave and a long silica terrace known as the Golden Fleece.

At the Agrodome, just outside Rotorua, sheep are the topic. An entertaining show takes visitors through aspects of the industry from lambing (*below left*) to shearing displays (*below*). The Agrodome was founded by the Bowen brothers, who consistently won world titles for speed shearing.

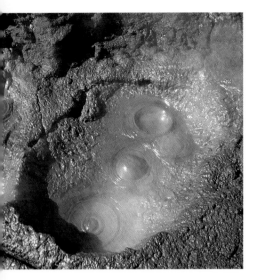

The earth's crust is riddled with cracks and holes at Rotorua, and the thermal splendours are legion. Boiling mud (*above*) and the Rainbow Terrace (*right*) at Orakei Korako. Silica terraces at Whakarewarewa (*below*) with Pohutu Geyser behind.

Water spurts through a lava tunnel at the eastern end of Lake Tarawera, yielding the lovely Tarawera Falls (*left*). Traditional Maori skills, including carving (*above*) are on display at the Maori Arts and Crafts Institute at Whakarewarewa. Arawa Maori (*below*) with a tribal waka on Lake Rotoiti.

CENTRAL PLATEAU

A harsh volcanic regime shaped the eerie central uplands of the North Island. Ancient cataclysms from Lake Taupo once overwhelmed the region, and the volcanic mountains Tongariro, Ngauruhoe and Ruapehu still smoulder. Yet the Central Plateau attracts visitors by the thousand: trampers and skiers to the dramatic walks and snowy slopes of Tongariro National Park, trout-fishing enthusiasts to its famous lakes and rivers.

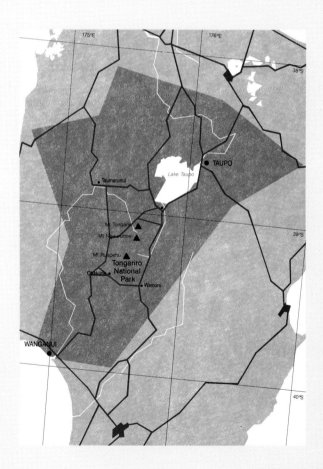

Above: *Pukekaikiore with Ngauruhoe in cloud.*
Left: *The three volcanoes of the Central Plateau, front to back, are Tongariro, Ngauruhoe and Ruapehu.*

The North Island's highest mountain, Ruapehu, though volcanic, lacks the classic volcano shape. In a 200,000-year history, the main summit vents have changed and the crest is a complex pattern of plateaus and ridges. Its most active vent now lies beneath the simmering and sulphurous 17-hectare expanse of crater lake (*right*). Ngauruhoe's more perfect symmetry (*below*) marks a volcanic cone perhaps as young as 2,000 years old, and a highly active one. The mountain erupts at intervals of between two and seven years. Behind Ngauruhoe, Mount Tongariro is the revered elder of the Central Plateau chain, battlescarred by eruption and weathering that extends back 500,000 years.

I n AD 186 a vast cloud darkened the earth: Roman records noted stars shining in the day, and the Chinese described a blood-red sun. In another hemisphere, right where today's anglers cast for their version of the big one, Taupo had erupted.

Taupo is New Zealand's biggest lake. At 600 square kilometres, it is also one of the world's largest volcanic craters. Its first eruption, some 300,000 years ago, blew out about 1,000 cubic kilometres of tephra, one of the mightiest explosions known, sending a wave of pumice as far as the Bombay Hills, near Auckland in the north, and forming the Kaiangaroa Plains in the east. The later eruption noted by Roman and Chinese historians, though far smaller that the first, still blasted some 100 cubic kilometres of material skyward – far overtopping the eruption of Mount St Helens in 1980 (1.7 cubic kilometres) and Krakatoa in 1883 (18 cubic kilometres).

Mounts Tongariro (1,968 metres), Ngauruhoe (2,290 metres) and Ruapehu (2,796 metres) are aligned southwest of Taupo on the same volcanic belt. All of them are still active: Tongariro and Ruapehu ancient, blast-scarred, and now only simmering, Ngauruhoe younger, still a pure cone, with a plume of steam and gas billowing frequently from the summit.

In 1887 the paramount chief of the Tuwharetoa tribe, Te Heuheu Tukino IV, gave much of the mountain chain to the nation. The gift laid the basis for the Tongariro National Park, some 78,651 dramatic hectares where trampers can move close to the steam, the foundry smells and the occasional rumble of a landscape still in the making. Highlights, requiring some tramping ability, are the fascinating mix of blowholes, mud pools and hot springs at Ketetahi Springs on the flanks of Tongariro; the crater of Ngauruhoe; and the hot acid expanse of Ruapehu's crater lake. The North Island's two major ski-fields are sited on Ruapehu. Whakapapa, on the northern slopes, has a motorcamp, shops and the grand Chateau Tongariro hotel at its base. On the south side, the Turoa field is accessed through Ohakune.

In winter, Ruapehu's mighty slopes provide the North Island's best skiing. Backed by the serrated skyline of Pinnacle Ridge, skiers relax (*above*) at the Whakapapa ski-field cafeteria, or ride to the top of the first chairlift (*left*).

The Central Plateau's reputation as a trout-fishing paradise brings anglers by the thousand into the region. A fisherman in the Tongariro River (*above*) tries his luck. The Queen Elizabeth II Memorial Army Museum (*right*) is the public face of the 3,000-strong New Zealand Army base at Waiouru. The museum displays in full-scale dioramas, realistic even down to the sweat on a soldier's brow, scenes from the campaigns fought by New Zealand troops. The displays are supplemented by print, photographic and audio-visual presentations. Taupo (*below*) is the country's largest lake, a 600-square-kilometre expanse popular for boating, fishing and swimming.

An apparently benign host to skiers, trampers and climbers, Ruapehu is carefully monitored for any signs of unusual activity. In 1953 a lahar from the crater lake swept away a railway bridge from the path of a passenger train at Tangiwai. Amidst the churning waters 151 people died. Early warning is sought for eruptions too. Though small over recent decades, the largest of them in 1975 blew out the crater lake. And as the North Island's highest peak, Ruapehu can generate weather dangerous to climbers.

The plateau is sparsely populated. Its main town, Taupo (pop. 15,000), dates from 1869, when it was a base for Armed Constabulary hunting the rebel Maori leader Te Kooti. The northern approach to today's town is dramatic. Power stations at Ohaaki and Wairakei tap geothermal fields adjacent to the Waikato River, and Wairakei's bores eject steam alongside Highway 1. The traveller may divert from the road to a lookout on trembling ground above the awesome Huka Falls, before the road south opens up a panorama of distant mountains, the lake and township below. Taupo has thermal pools and is primarily a holiday resort, particularly for anglers. Its fishing season is open throughout the year, and the size and frequency of the catches are a byword on the international trout-fishing circuit.

The bays and rivers of the lake's eastern edge are dotted with fishing settlements round to Turangi on the southern shore. This township was formerly a base for tunnelers on the Tongariro hydro-electric power scheme, and now accommodates anglers and skiers. Nearby, the older village of Tokaanu offers hot springs. From both towns there is access to popular fishing spots on the Tongariro River. Highway 1 then crosses the Rangipo Desert, with views of the three volcanoes, to reach Waiouru, New Zealand's largest army base, on the southern edge of the plateau.

Swans and ducks on Lake Taupo (*above*). Numerous camping spots and motel units around the lake shore cater for a summer influx of tens of thousands of visitors, yet beaches around the lake shore are plentiful and retain an uncrowded feel (*below*).

A distant Lake Taupo, seen from the Waitahanui Hills (*above*). At historic Huka village, 4 kilometres north of Taupo, (*right*), the past has been reassembled with Victorian buildings, pioneer artifacts and horse and buggy rides. The Huka Falls (*below*) have a modest 11-metre drop, but no waterfall in New Zealand can rival the impression of raw power as the Waikato River leaves Lake Taupo to squeeze through a narrow cleft and thunder over that chasm.

West of the plateau the land falls away in crumpled country, much of it covered in pine forest or native bush. The Main Trunk railway line pushed through here in the early 1900s, giving economic life to the town of Taumarunui, and, further south, Taihape.

In these western hills the Wanganui River begins a 290-kilometre journey, much of it through the protected scenery of the Whanganui National Park to the sea. The river is ineluctably beautiful and offers the canoeist or small-craft owner a run of some 210 kilometres downriver from Taumarunui to the coastal city of Wanganui (pop. 41,000). For those prepared to brave the rapids, the top part of this run, down to Pipiriki, opens some of New Zealand's most beautiful interior bushlands to view. Below this small settlement the river broadens and flows on past Jerusalem and the grave of James K. Baxter, best-loved of New Zealand's contemporary poets, finally to the sea.

Wanganui (pop. 40,000) takes its name and a gentle charm from the river. It is one of New Zealand's oldest cities, the centre of a fertile farming area, and a departure point for upriver exploration by road, jetboat or restored river boat.

The deep roar of harnessed geothermal energy surrounds the Wairakei power station (*below*) as it converts the earth's superheated steam to electricity. The station's 190,000-kilowatt capacity already make it, of its type, a world leader, but New Zealand engineers calculate that geothermal fields in the region can yield another 500 megawatts of power.

Sheep in a holding pen near Taumarunui await the shears (*above*). The Tangarakau Gorge (*right*), a beauty spot southwest of Taumarunui on Highway 43, sited amidst some of the North Island's most rugged bush country. The long and lovely Wanganui River (*below*), seen from the crest of Aramoana.

Wanganui City (*left*) began in 1840 as a New Zealand Company settlement called Petre. A dubious land transaction blighted the new town's beginnings, leading to skirmishes with local Maori and some deaths. Wanganui's long history since, as a centre to rich surrounding farmland, has given it unusual poise and a strong arts traditions. The domed Sarjeant Art Gallery has a good New Zealand collection, the museum features a large Maori collection, and the Four Seasons Theatre offers plays. The Virginia Lake Reserve (*below*), with its winter gardens and walk-through aviary, is one of the city's beauty spots.

EAST CAPE

At Kaiti Beach, near where Gisborne city now stands, James Cook first set foot on New Zealand, ending a thousand years of Maori isolation. First-visited the great cape may have been, yet it has stayed distinctly Maori, its character described by a proud tribal motto:

Ko te Waiapu te awa,   *Waiapu is the river*
Ko Hikurangi te maunga,   *Hikurangi is the mountain,*
Ko Ngati Porou te iwi.   *Ngati Porou are the people.*

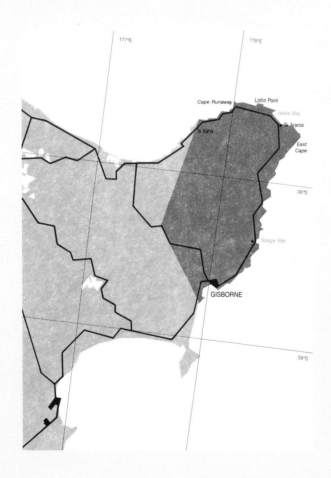

Above: *Gisborne.*
Left: *Cape Runaway.*

The tekoteko, or gable figure, stands sentinel on the Te Kaha Whare Runanga, or meeting house, at Te Kaha (*above*). Stark sentinel also to past sacrifice is the war memorial on the marae, the square of land in front of the meeting house that is the spiritual heart of the tribe. The western coast of the East Cape has many good beaches, and the inlets of Whanarua Bay (*right*) are amongst the prettiest of them. Lottin Point (*below*) is the most northerly part of the cape accessible by road.

The East Cape is the North Island's most remote region. Only one road traverses it, the scenic Highway 2 from Opotiki past river and forest views in the Waioeka Gorge to Gisborne.

A second route, Highway 35, goes around the cape, double the distance but dazzlingly beautiful. Beyond Opotiki the volcanic White Island plumes upon a sea horizon. Deserted beaches open up at the roadside. The distinctive richness of the East Cape Maori carving style adorns the War Memorial Hall at Omarumutu and meeting houses at other settlements on the western shore.

Inland, hills of the formidable Raukumara Range crowd down to the sea. The Raukumara Forest Park encompasses much of the interior, but access is difficult. At Maraenui Beach, the road traveller passes over the Motu, narrow-gorged and fast-flowing still at its mouth, the fiercest and finest of all New Zealand's rafting rivers.

The road turns inland just before Cape Runaway, named by Cook in 1769 for a party of Maori who canoed out to the *Endeavour*, but headed rapidly for the shore after Cook sent cannon-fire screeching overhead.

Hicks Bay is the first of a succession of settlements on the eastern coast where decaying wharves are testament to a vanished sea trade. The tough little steamers of Richardson and Co. used to call here, as they did all the way up the east coast. The skills of the crews on a harbourless shore included running wool bales out to the ships by surfboat. The service stopped in the 1950s, and the freezing works also closed. The Tuwhakairiora meeting house at Hicks Bay is one of the finest on the coast.

On the road to East Cape, looking northwest across Hicks Bay towards Matakaoa Point (*above*), and the same coastline (*below*), looking northwest from the cape. In this isolated region, New Zealand's most famous teacher, Sylvia Ashton-Warner, pioneered putting Maori words back in the teaching syllabus.

East Cape (*above top*) and St Mary's Church at Tikitiki (*above and below*). The church was built in 1924 to honour Ngati Porou war dead. Its interior tells tribal history, its stained glass depicts fallen soldiers.

Te Araroa has a thirty-year-old pub alongside a 600-year-old pohutukawa tree, said to be the country's oldest. Fighting pa once stood on the summits of the surrounding hills, for Te Araroa marks the northern boundary of Ngati Porou territory. From the town a gravel road leads along the shoreline to East Cape, New Zealand's most easterly point, climbable by 707 steps to the lighthouse there.

The main road heads inland, then crosses the Waiapu River, whose valley opens up a strangely discordant landscape through to the great double crag of Hikurangi (1,753 metres). The mountain is the East Cape's highest and, as dawn rolls onto New Zealand, Hikurangi is the first mainland tip to catch the sun.

The Waiapu Valley is the site of the major Ngati Porou towns of Tikitiki (with its beautiful Maori church of St Mary's), Waiomatatini and Ruatoria. In the 1860s, Ngati Porou leaders raised war parties against the anti-government Pai Marire (Hauhau) movement and the rebel army of Te Kooti, which had massacred 70 in a single swoop at Poverty Bay. Their actions kept Ngati Porou land safe from the confiscations suffered by Maori tribes elsewhere. Much of the steep, sheep-grazed land and river flats are still Maori-owned, and most of the rural people are Maori. Ngati Porou tribal identity is strong: their Apirana Ngata is known as perhaps the greatest Maori politician of the first half of the twentieth century, and their fighting men are famous for deeds of bravery with the Maori Battalion in World War II.

Hot springs and natural-gas emissions are the main attraction at Te Puia, before the road returns to the coast at lovely Tokomaru Bay. Once relatively prosperous with a freezing works, and a new wharf completed in 1940, the town is quiet now, with both wharf and works derelict.

Nearby Anaura Bay recalls the first peaceful encounter between Maori and European.

New Zealand's history of contact had not been, until then, auspicious. In 1642 Abel Tasman lost four men to a Maori attack on a long-boat and never set foot ashore. On 6 October 1769, Nicholas Young, a twelve-year-old boy aboard the *Endeavour*, sighted the New Zealand coast from his masthead perch. Young Nicks Head, the southern promontory of Poverty Bay, is named for him. But Cook's landing at Kaiti Beach on 8 October was marred by misunderstandings that left one Maori shot dead on day one, and another five killed by British musketfire on day two. Cook left without provisioning his vessel, and sourly named what is in fact a fertile region, Poverty Bay.

At Anaura Bay, though, gifts were exchanged and courtesy prevailed. After two days ashore, Cook and his men were directed south to Tolaga Bay, an anchorage protected by Pourewa Island. There the ship's artist, Sydney Parkinson, waxed lyrical over a natural arch that still exists. The bay is now a farming centre, too close to Gisborne to have developed into a market town.

Gisborne (pop. 33,000) is the city centre to the Poverty Bay region. The rich surrounding plains produce some of New Zealand's best wines, and sheep farming flourishes on the hills behind. The city is also a fishing port and food-processing centre. Cook is its emblem. Metal models of the *Endeavour* sail up the main street, affixed to lamp-posts, and a statue of the famous navigator stands upon the town's highest lookout, Kaiti Hill.

The horse and its Maori rider (*above*) are still a part of daily life on the cape. Gisborne (*below*) is the most remote North Island city, its rich flatlands enclosed by rugged hills and a rolling surf.

HAWKE'S BAY

Hawke's Bay's northern hinterland is cloaked by a forest fastness, Lake Waikaremoana its clear centre, and the Tuhoe Maori its traditional people. South from there the change from wilderness to civic grace could not be more marked. Destroyed by earthquake in 1931, the city of Napier has re-created itself as one of New Zealand's most stylish towns.

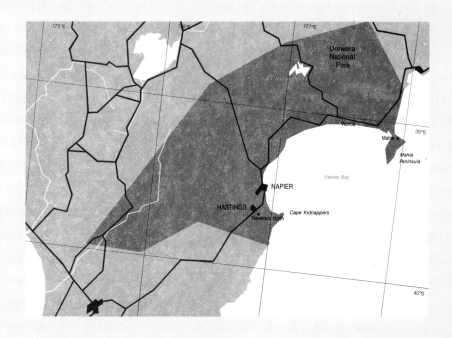

Above: *Whakaari Bluff, Hawke Bay.*
Left: *Vineyards near Napier.*

The Papakorito Falls (*above*), in the Urewera National Park, plunge 18 metres over an old fault in the Aniwaniwa Valley. The most characteristic feature of the park, though, are its forests (*right*). The park preserves the largest area of native forest still remaining in the North Island. Much of the forest is beech, producing a high canopy that delicately filters light onto a forest floor carpeted in moss and ferns. Elsewhere tawa dominates, a mightier tree whose berries attract the beautiful native woodpigeon. The park's bird-life is prolific, with tuis, kakas, parakeets, bellbirds, tomtits, fantails, grey warblers, bush robins and riflemen all common.
Opposite: Lake Waikaremoana (*above and centre*), seen with the dramatic Panekiri Bluffs in the background, is the park's blue centrepiece. The lake is the headwater of the Waikaretaheke River, a tributary to the broad Wairoa River (*below*), which flows calmly past Wairoa township to the sea.

84

M ahia Peninsula was known to the Maori as such a frequent stranding zone of whales that a magical lure, or mauri, was said to be buried in its hills. The whales were tapu then, but not to later whaling stations, which all but eliminated the southern right whale from these waters. Modern Mahia is an isolated but favoured camping spot, serviced by a single township.

The peninsula forms the northern tip of Hawke Bay, as good a place as any to contemplate not just the single sweep of the bay's shingle shore, some 100 kilometres to Napier (pop. 53,000) in the south, but also to understand the grammar of the place: Hawke's Bay is the name of the province, Hawke Bay the name of the bight.

Highway 2 skirts the bay, leading first to Wairoa, the main township of northern Hawke's Bay. It is a market centre to surrounding sheeplands, distinctive for its slow, wide river and the lighthouse, which signalled once from Portland Island on the tip of the Mahia Peninsula and now stands mute on the main street. Inland from there on Highway 38 is the Urewera National Park and its lovely centrepiece, Lake Waikaremoana. The lake, one of New Zealand's most beautiful, was created 3,000 years ago when a great landslip closed the river gorge between the Panekiri Bluffs and the Ngamoko Range.

Other highlights of the national park are nearby: waterfalls and a bush walk to Lake Waikareiti, which, like Chinese boxes, encloses an island that itself encloses a small lake.

The Rothmans Building (*right*) in Bridge Street, Napier, is a good example of the architectural panache with which a city destroyed by earthquake and fire in 1931 resurrected itself. A major attraction on Napier's beautiful Marine Parade are the dolphins of Marineland (*below*), who perform regular shows, but the Norfolk pine-fringed parade has museum displays, an aquarium, a nocturnal wildlife centre, a skating rink, sound-shell, a mini-golf course... in short, enough, and in sufficient variety, to keep every visitor amused.

The highway north wends onward through bush-clad wilderness. These territories were once the domain of the Tuhoe, an upland Maori tribe known as 'the children of the mist'. The road emerges at Ruatahuna, a Tuhoe town, then winds on to Murupara.

The floor of the Pacific Ocean is advancing upon New Zealand's east coast at an unstoppable five metres every ten years. Minor shakes may help accommodate that Pacific Plate as it planes downward beneath the country, but occasionally and inevitably, a major earthquake occurs.

At 10.47 a.m. on 3 February 1931, the ground beneath Napier city rose two metres. The city was destroyed by that convulsion and the following fires. In the city and surrounding areas some 250 people died. The gigantic seesaw dropped Wairoa, at the other end of the bight, downward, but raised 3,645 hectares of land from the seabed at Napier.

The city was rebuilt within a few years of the earthquake, and unwittingly gained an architectural legacy. Art Deco building was fashionable in the 1930s, and the style was seized upon by Napier architects because it used quake-proof reinforced-concrete construction. The city is a showplace of Art Deco buildings. Napier's setting, too, makes it one of New Zealand's most attractive small cities, with a soundshell, gardens and a marineland occupying its long foreshore.

On that same foreshore, between the Tutaekuri and Ngaruroro Rivers, is a tablet to commemorate the founder of the town, William Colenso, who occupied a mission station here in 1844.

Sunset on wetlands close to Napier (*below*) and Lake Tutira (*above*). This lovely sheet of willow-fringed water was inspirational to the late William Guthrie-Smith, a sheep farmer and a conservationist before people knew the word's meaning. In the 1920s he berated the 'rat-like pertinacity' by which local farmers destroyed New Zealand's distinct fauna and flora. He developed a large sheep station and still found time to publish contributions to New Zealand's natural and social history. The lake was declared a bird sanctuary at his instigation.

Top: Napier's foreshore, with its Esplanade marked out by Norfolk pines. The view from Bluff Hill looking northwest to snow on the Kaweka Range (*above*), and the port of Napier (*below*).

The Heretaunga Plain is New Zealand's greatest apple-growing region. Grapes also flourish, and the area has become an important wine producer. Hastings is a city of the plains, close to Napier but distinct in itself. Nearby is the country's oldest winemaking establishment, the Mission Vineyards at Greenmeadows.

Hawke Bay ends in the dramatic promontory known to the Maori as Te Matau a Maui – Maui's hook – an apt name, for if the bay is the mouth of Maui's fish, as legend has decreed, and if the fish was hauled from the sea by Maui, as tradition assures us, then this is exactly where you'd expect to find his hook. During Cook's adventurous first days on the coast of New Zealand, however, a young Tahitian on the *Endeavour* was grabbed from the bow-chains of the barque by a party of Maori. Cook opened fire on their canoe, and in the confusion the boy swam back to the ship, but at the cost of three Maori lives. Cook's name of Cape Kidnappers replaced the older one.

The cape is now a seasonal home to gannets. Ministering to nests just pecking distance apart, some 4,500 gannet pairs populate headlands high above the rolling surf. The breeding colony is one of New Zealand's great sights. The recommended time to visit is between early November, when young chicks are hatching, to late February, when dispersal to Australia and other destinations has begun. The sanctuary is closed from the beginning of July until mid-October.

Croplands outside Napier (*left*). The apple orchards of Havelock North (*above*) and the Heretaunga Plain are the nation's most extensive, and a major export earner. The walk to Cape Kidnappers (*below*) passes beneath dramatic cliffs, across the fossil-laden Black Reef, and finishes with a climb up to a lighthouse and beyond to a vantage point overlooking the Cape Kidnappers gannet colony. In its site and size, this is New Zealand's most spectacular colony.

WELLINGTON ·
WAIRARAPA · MANAWATU

The Wairarapa, on New Zealand's west coast, and the Manawatu-Horowhenua region on the east coast converge finally at the tip of the North Island, and the city of Wellington. With a near-perfect harbour set amidst steep hills it is a city subject to moody winds from Cook Strait, and one whose wetted finger is raised to the nation's moods also, for it is New Zealand's capital.

Above: *Paraparaumu Beach.*
Left: *Wellington high-rise.*

Hay bales on the fertile Makakahi River flats near Eketahuna (*above*) and a sheep muster near Dannevirke (*right*). Sheep have sustained the Wairarapa economy since the time Charles Bidwell drove a flock of 350 around the coast from Wellington in 1843 to establish New Zealand's first sheep station, at Wharekaka. Wairarapa sheep now number in the millions and are the dominant farming industry. The monolithic crag of Castle Rock (*below*) affords a wide view over the Wairarapa's most beautiful coastal haunt, Castlepoint.

The Wairarapa hills, which rise from a rugged east coast to the central Ruahine, Tararua and Rimutaka Ranges, are traditional sheep country. The country's first sheep station was established here in 1843, though the alluvial flats now support dairy farming, cropping and horticulture, and small population centres.

South from Hawke's Bay, Highway 2 reaches Dannevirke, a name meaning Danes' work, in tribute to Scandinavian pioneers who shaped much of nineteenth-century Wairarapa. Further south, at Woodville, is the Manawatu Gorge. The Manawatu River rises east of the Ruahine Range then cuts through it to discharge on the west coast, a geographic marvel naturally engineered by riverbed erosion over millions of years, matching the slow uplift of the land. Spectacular gorge walls, sometimes 300 metres above the river, are the result.

Masterton (pop. 20,000) is the region's main town. It gives access west to the Tararua Forest Park, and east to Castlepoint Beach, the lagoon there, and the craggy, 162-metre Castle Rock. Highway 2 goes on past Carterton, Greytown, Featherston and the shallow Lake Wairarapa to cross the Rimutaka Range to Wellington. Highway 53 diverges at Featherston to head coastward. The country grows more rough but yields such scenic beauties as the weathered conglomerate spires of the Putangirua Pinnacles. A rough road leads on to Cape Palliser, southernmost point of the North Island.

A shepherd and his dog muster sheep on the lonely Wairarapa coast (*above*). A mere 46,000 people live amidst Wairarapa's wide spaces, a low figure even by rural standards. Castlepoint lighthouse (*below*) is one of the country's tallest. It stands on the northern arm of the long limestone reef that provides sheltered beach swimming and a lagoon at this favourite holiday settlement.

Wellington city and harbour at night (*top*). The houses that colonise in unruly ranks Wellington's steep slopes are mostly wooden and often of Victorian or Edwardian vintage. At Oriental Bay (*above*), a beach suburb, fashionable dwellings jostle for space on a steep rim of the inner harbour. Those above The Terrace (*right*) overlook the city centre's high-rise. The intimate conjunction of inner-city suburbs and city centre is part of Wellington's character.

The Manawatu-Horowhenua region boasts the North Island's most extensive river flats. The plain is trisected by the Rangitikei and Manawatu Rivers and bisected regularly by Skyhawk fighter planes returning to their base at Ohakea. Otherwise little disturbs a shimmering farmland that extends from the distant spine of the Tararua Range to a pleasant, sandy coastline.

Palmerston North (pop. 67,000) is the only city on the plains. Levin, further south, is a prosperous farming and market-gardening town. At Otaki is the grave, reportedly empty, of the feared Maori chief Te Rauparaha. From a base on nearby Kapiti Island, the Ngati Toa leader launched musket-armed war parties on the South Island. During the 1820s and 1830s, he and his Te Atiawa allies struck as far south as Cannibal Bay in Southland, terrorising Ngai Tahu and other tribal settlements the length of the coast.

Over one thousand years ago, according to Maori tradition, two canoes under the command of the Polynesian explorer Kupe sailed into the great circular harbour at the tip of Te Ika a Maui. The place names Kupe gave to the harbour's main features are still known, suggesting the memory of this first visit was perhaps kept alive by a small colonising group. If so, Wellington, which is ritually spurned by much of New Zealand as a fit place to live, was one of its first settlement sites.

More definite history begins in the twelfth century with colonisation by a Hawke's Bay subtribe, Ngai Tara. Descendants of this, and other Wairarapa tribes, watched the arrival into the harbour, on 2 November 1773, of the *Resolution,* Cook's vessel on his second of three expeditions to New Zealand. Cook didn't explore the harbour. It was left to Captain James Herd of the *Rosanna* to chart it in 1826, name it Port Nicholson and report: 'The navies of all the nations of the world could lie at anchor here.'

The square-rigged brigatine *Spirit of Adventure* on a training voyage at the mouth of Wellington Harbour (*above*). Jan Morris, the travel writer who set out to visit every major city of the world, wrote that Wellington was 'unexpectedly ennobled, like San Francisco, by that queen of the tramlines, the cable car'. The car (*below*) runs a regular shuttle between Lambton Quay and Kelburn.

PARLIAMENT GROUNDS

Wellington is New Zealand's capital city. The Beehive (*right*) houses the Prime Minister's office and the Government Executive. A visiting English architect, Sir Basil Spence, first sketched the design on a restaurant napkin, apparently inspired by the beehive symbol on a New Zealand matchbox. The building was opened in 1981. The bronze statue is of Richard John Seddon (King Dick), Liberal Premier from 1893 to 1906. The adjacent Old Parliament Building (*below*) was built in 1922 of Coromandel granite and Takaka marble. It houses the debating chamber for the House of Representatives. The wonderfully gothic General Assembly Library (*above*), also in Parliament Grounds, is at the service of members of Parliament and has the country's most comprehensive collection of New Zealand publications.

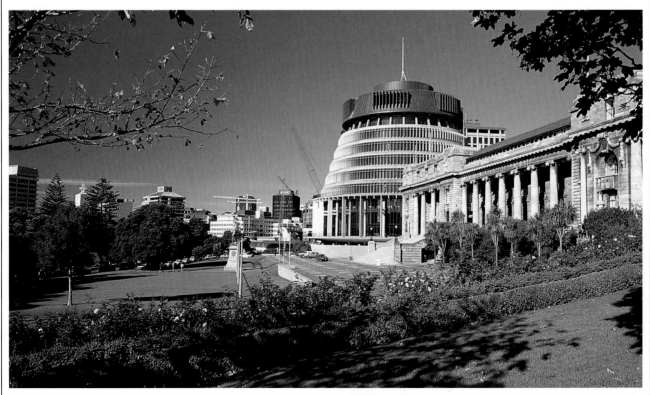

The first white settlers arrived aboard the *Tory* in 1839, a New Zealand Company expedition organised from Britain by Edward Gibbon Wakefield with the support of the Waterloo hero the Duke of Wellington, for whom the new settlement was named. With the formal annexation of New Zealand by Britain in 1840, Wellington pressed its claim to be capital. Lieutenant-Governor Hobson chose Auckland, but by the 1850s, disgruntled southern politicians were travelling weeks to each session of the northern parliament. In 1865 the House of Representatives was shifted closer to the country's geographic centre – to Wellington, which became the capital.

Wellington and the surrounding urban areas have a population of 328,000. The north end of the city is dominated by the buildings of government. The concentric rise of the Sir Basil Spence-designed 'Beehive' houses the Executive, with Parliament Building and the General Assembly Library adjoining it. Nearby are the National Library of New Zealand; the Government Building, opened in 1876 to contain the entire civil service and still the biggest wooden building in the Southern Hemisphere; and the old St Paul's Cathedral, built in 1866.

South from there, high-rise buildings stand in a cliff-like mass along the old foreshore of Lambton Quay, or colonise the newer lip of harbour reclamation. Behind the business centre, a cable car, serpentine roads and zig-zig paths link the inner suburbs to the city's heart. Excepting only the Hutt River Valley to the north, the harbour is surrounded by steep hills, and the predominantly old villas practically overhang its shores, bound to the land by concrete retaining walls, and hope.

The hope is against catastrophic earthquake. The city is built on a major faultline, and creaks occasionally to small shakes, insignificant compared with the quake of 1855, when the fault suddenly sheared nearly three metres. Wellingtonians look across their harbour to winter snow on the Rimutaka Range, a picture-postcard prettiness edged with dread that the range is, in fits and starts, one of the most rapidly rising blocks of land in the country.

The citizens have also learned to love, or at least to lean into, their famous winds. The prevailing New Zealand westerly is amplified by Cook Strait into a bustling nor'wester, which bounces in frequent thirty-knot gusts around the city streets. Wellingtonians call it bracing, but when it blows the other way, even seasoned citizens scurry before a southerly buster. In the calms between, though, there is not a city in New Zealand to rival Wellington's clarity of air, its compact sense of self, and the encircling beauty of its setting on the North Island's southern tip.

Mirror glass and an impressive architectual style (*above*). Wellington's building space is constricted and the city is head office not just for the Civil Service but for big business. Prestige is important, and the new architecture has a more studied air of quality than other New Zealand cities. The city's outskirts are still barely tamed, and edged by a rugged coast (*below*).

The view from Paekakariki Hill, (*above*), looking northwest towards the seaside settlements of Paekakariki and Raumati. The Levin cattleyards (*right*). The town is a prosperous market garden and farming centre. Its extensive surrounding land was first opened to city markets by the arrival of the railway in 1899. The town was therefore named for William Levin, a director of the company that laid the connecting rail-line.

The Manawatu's rich alluvial plains have been laid down by two major rivers; the rich mixed-crop land near Marton (*left*) by the Rangitikei. On the banks of the same river stands the Royal New Zealand Air Force base at Ohakea, an RNZAF Skyhawk fighter seen here under youthful inspection (*above*). Placidly, through river flats and past cliffs of its own making, winds the Rangitikei itself (*below*).

MARLBOROUGH·
KAIKOURA

The quiet bays and protected waterways of the Marlborough Sounds have charmed voyagers from James Cook to today's inter-island ferry travellers. Beyond are the sun-soaked river valleys that make the region's wines famous, and the first hint of South Island snow as the Kaikoura Ranges crowd to the sea.

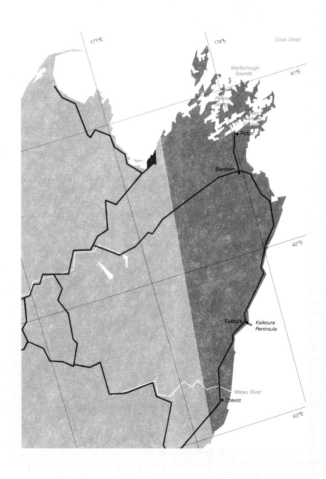

Above: *Elaine Bay.*
Left: *Marlborough Sounds.*

The inter-island ferry links New Zealand's North and South Islands, shuttling railway wagons, road vehicles and passengers across Cook Strait four times a day. The crossings take three and a half hours. The ferry sails into Tory Channel (*above*) and approaches its South Island terminal at Picton (*right*). The town's marina (*below*) is haven to small craft that ply the Sounds. Road linkages within the Sounds are sparse. A boat or a six-hour tramp provide the only access to historic spots like Ship Cove.

K upe's mighty shadow hovers over the Marlborough Sounds. Today's inter-island ferry passes two islands known as The Brothers, but the Maori called them Nga Whatu-Kaiponu, 'the eyeballs that won't let go'. In legend they were the last glare of a dying octopus slain by Kupe. As the inter-island traveller enters Tory Channel, the legend is manifest. The petrified tentacles of the slain giant lie all around.

Another myth explains the intricate filigree of land and water as the carved prow of a supernatural canoe wrecked here. The tale echoes the findings of modern geology, that the Sounds are a series of drowned valleys, tilted downward over millenia by New Zealand's main mountain-building force, the Kaikoura Orogeny.

The two main waterways are Queen Charlotte Sound, named by Cook, and Pelorus Sound. Cook arrived in 1770 to careen his vessel at Ship Cove. He first saw the strait named after him when he climbed nearby Arapawa Island.

The South Island terminal for the inter-island ferry is at Picton, near the base of Queen Charlotte Sound. The port bustles with the discharge and loading of inter-island railway wagons, cars and travellers. Behind, the town has a pleasant Victorian charm. From its pretty waterfront, launches ply to hideaway guest houses and fishing spots within Queen Charlotte Sound.

Tawhitinui Reach (*below*), a part of Tennyson Inlet, seen from the long road out to French Pass. These western reaches of the Sounds remained unexplored by Europeans until as late as 1827, when the French explorer Dumont d'Urville first charted them.

Though the Wairau Plains are reckoned to have near-perfect conditions for the production of white wines, viticulture did not begin here until 1973. Montana's vineyards (*right*) were amongst the first planted, and their exported white wines quickly made Marlborough New Zealand's best-known wine region. Blenheim's horticulturally based prosperity is reflected in the town's well-groomed municipal gardens (*below*).

A road runs northeast from Picton to Port Underwood, formerly a whaling station on the eastern flank of the Sounds. To the northwest, the town of Havelock serves Pelorus Sound, and a secondary road winds out along the tortuous peninsula dividing the two main waterways. Beyond Havelock, through the Rai Valley, is road access to beautiful Tennyson Inlet, surrounded by the largest of the Marlborough Sounds Maritime Park's hundred-odd scenic reserves. Historic French Pass is reached by the same route. There, in 1827, the French explorer Dumont d'Urville risked his ship to run through a narrow, reef-encrusted channel. In the same waters the dolphin Pelorus Jack, for twenty-four years from 1888, met the ships that ran through to Nelson.

South of Picton on State Highway 1 is Blenheim (pop. 23,000), Marlborough's main town. The surrounding plains and the Wairau River valley are that potent horticultural mix of rich soil and steady sunshine – the region often leads the country in sunshine hours. Cherry orchards abound, all the stone and pip fruits too, but wine is the elixir. The Wairau Plains have the most extensive vineyards in the country, producing white wines, and with producers Champagne Deutz and Veuve Cliquot now in residence, sparkling wines par excellence.

Visible from the highway beyond Blenheim is the Wairau Boulder Bank. Though still not open to the public, it has provided the best evidences to date of so-called moa-hunters, the peaceable Polynesian immigrants who populated New Zealand in mediaeval times. They drove the flightless moa – the largest bird ever known – into such culs-de-sac as the water-enclosed Wairau Bar and slaughtered them for food.

State Highway 1 rolls on across loess hills to an unusual double-decked road-rail bridge at the Awatere River valley, and the farming settlements of Seddon and Ward. Upriver from there lies Molesworth Station, a 1,800-square-kilometre sheep and cattle run, New Zealand's largest, administered by Landcorp Farming Ltd, a state-owned enterprise.

The Wairau Plains are New Zealand's sunniest ground, a horticultural and agricultural El Dorado, vibrant with deep colour and, by summer's end, tanned brown by the heat. Haystacks on the plains (*below*).

The evaporation ponds of Lake Grassmere, at the base of Cape Campbell, produce New Zealand's salt — the last industry in sight before the road joins a lonely and beautiful coast. In 1978, air-traffic controllers at Blenheim's Woodbourne aerodrome watched four unidentified flying objects hover off this coast. The visual sightings were confirmed by radar at Wellington airport, and by a transport plane flying down the coast. Ten days later a television team from Australia flew the same route, on a routine wrap-up of that story, and had an unexpected close encounter. The United States National Investigations Committee on Aerial Phenomena later endorsed their film as showing a genuine UFO — the first such endorsement ever given by NICAP in its investigation of some 22,000 UFO sightings.

Beyond the Clarence River the flat-lands disappear. The scenic highway is notched upon headlands, the railway disappears in a series of tunnels as the black crags of the Seaward Kaikoura Range meet the sea.

Kaikoura township has the salty character of an old fishing settlement. The Maori fished here first. Kai, the beginning of the name, means food. They gathered it from craggy offshore reefs, the same food the fishing fleet here gathers still, the most succulent crustacean of all — its Maori name koura, in English, crayfish. The town's long history includes the depredations of a whaling station, its artefacts still preserved on the shoreline and the local museum. The great mammals still sport off the coast, and when they do, small craft set out from Kaikoura on whale-watching expeditions. A walk to the peninsula tip is another natural restorative. A colony of seals basks there.

South of Kaikoura, before Highway 1 reaches Canterbury, Highway 70 diverges west to cross the Lewis Pass into Westland. Midway to the pass, in a natural amphitheatre of inland plain and surrounding mountain, is Hanmer Springs, a spa town. At 360 metres above sea level, Hanmer's climate is cool, its pools hot, its hotels, dating from a time when thousands chased the vapours of thermal cures, historic.

Kaikoura's many offshore reefs are a perfect habitat for crayfish (*top*). The town's buildings retain the century-old character of an ex-whaling town (*centre*). A seal colony, protected now as are the whales from man's depredation, basks at the tip of the Kaikoua Peninsula (*above*). As mountain ranges shoulder right to the sea, State Highway 1 tunnels through a headland (*right*) and sweeps on above a rugged and foaming coastline.

Sheep on the riverflats near Cheviot (*left*). The Waiau River bears shingle from its alpine headwaters to the inland amphitheatre of the Hanmer Plains (*below*), 360 metres above sea level. The river reaches the coast just north of Cheviot, and State Highway 1 crossing there provides the traveller with a first glimpse of a South Island 'braided' river.

CANTERBURY

The snowy Southern Alps deckle-edge the western horizon of New Zealand's largest province. At their feet spreads an immense piedmont apron, the Canterbury Plains. The grid of grain and sheep-farming fields there gives the South Island its agricultural heart and sustains its largest city, Christchurch.

Above: *Christchurch.*
Left: *Lake Pearson.*

Between Porters Pass and Arthur's Pass, on State Highway 73, lie the limestone outcrops of Castle Hill (*top*). The place was once a staging post on the coach route to the West Coast. The same area (*centre*), looking west toward the Craigieburn Range. The mountains are winter host to skiers on the Broken River and the popular Craigieburn Valley ski-fields. The headwaters of the Waimakariri River (*right*) lie amidst mountains of the Arthur's Pass National Park. The river is braided in unthreatening channels as it begins its journey to the sea, but the broad gravel bed hints at a mightier flow. During flood, the separate channels are lost amidst the raging force of a river over 100 metres wide, rumbling with bed-rolled boulders and shifting shingle. Such river power built the Canterbury Plains.

110

T he mighty Southern Alps are New Zealand's most magnetic landscape and, geologically, its most awesome. Along the Alpine Fault the slow uplift of rock over the past 20 million years has been estimated at fourteen kilometres. Erosion kept pace for much of that time, and only over the past two million years has the uplift increased sufficiently for the Alps to emerge as a lofty mountain range.

Eroded rock, spread eastward by the action of glaciers and huge, braided rivers like the Waimakariri and Rakaia, built the Canterbury Plains.

Another prominent landscape feature is the two ancient coalesced volcanoes that make up Banks Peninsula. Originally an island off the coast, they were joined to the mainland perhaps only 10,000 years ago, by the eastward creep of the plain.

Dinornis maximus, the biggest of the grazing moas, once stalked the plain, to be extinguished later by the first Polynesian immigrants, the moa-hunters. The Ngati Mamoe, then the Ngai Tahu followed, but in the 1830s fell victim to bloody raids by the North Island's Ngati Toa and Te Atiawa. In 1848 the government purchased eight million hectares of land from the Ngai Tahu for £2,000.

The province was colonised by the Edward Gibbon Wakefield-inspired Canterbury Association. Its aim was to remake a tussocky wilderness in the image of pastoral England – an upper-class gentry and Anglican clergy in charge of land and spirit, with labour provided by an underclass of servants and labourers.

Christchurch from the Port Hills, looking north (*below*). The Ferrymead Estuary was once a port to shallow-draught vessels but fell into disuse from 1867 when the tunnel through the Port Hills gave Christchurch an easier access to Lyttelton's deep-water harbour.

Known as the most English of New
Zealand's cities, Christchurch has its
own town crier (*above*). The Avon
(*right*), home to ducks and punts,
trailed by willows and spanned by
Victorian bridges, conjures more
memories of old Albion. The
Christchurch Cathedral (*below*) has
presided over the city's central
Cathedral Square since 1874.

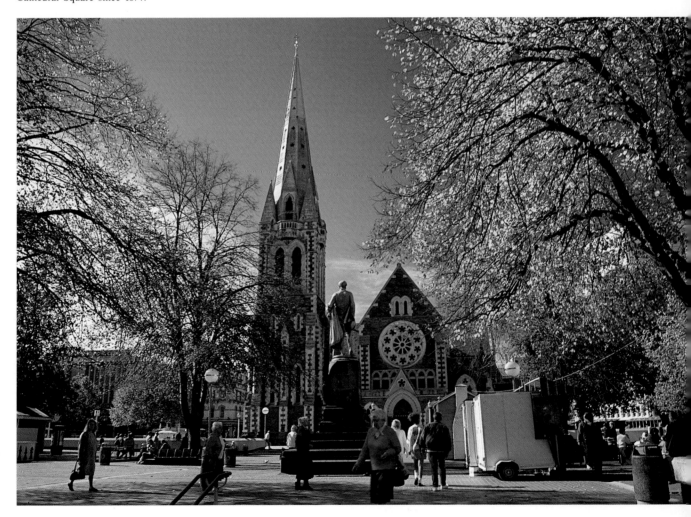

Canterbury took its own course. No markets existed for the products of an intensive English-style agriculture. The only profit was in sheep. The plains and the rolling foothills of the Alps became a big man's frontier of sheep runs so large their boundaries were marked out by riverbed and mountain range. The runs were broken up later into smallholdings, and the independent arable farmers of the plains were a further violation of the original dream. Yet Canterbury's founding ideals still show. A patrician style still characterises the high-country sheep runs, and there is still merit in being a descendant of the founding 'pilgrims' who arrived in the first four ships of 1850. Christchurch itself (pop. 300,000) is the most English of the New Zealand cities.

The city's centre is the square where New Zealand's most graceful Anglican cathedral presides. Around that, tree-lined streets have the formal regularity of a city planned on flat land. The Avon River is a meandering counterpoint. Its punts and ducks, its bank-side daffodils and fountains, its pretty over-arching bridges give Christchurch much of its charm. So too does Hagley Park with the Botanic Gardens at its heart, the largest city park in New Zealand. Christchurch is also justly proud of its many old stone buildings. The Gothic style predominates, and is best seen in the buildings of the old university, now an arts centre, and in New Zealand's only surviving provincial council chamber.

The city's port is Lyttelton, a steep-sided town quite distinct from Christchurch, reached by tunnel through the Port Hills, or by a more scenic Summit Road. Lyttelton Harbour is the sea-flooded crater of an ancient volcano.

The main road to Banks Peninsula runs past Lake Ellesmere, a shallow, 20,000-hectare lagoon separated from the ocean by a shingle spit, and an important wetland supporting thousands of wading birds and swans. Beyond the lip of the adjacent hills is Akaroa Harbour – another sea-flooded crater. Much of the surrounding land was bought by the captain of a French whaler in 1838. By the time French colonists arrived in 1840, though, New Zealand was already a British colony. The French immigrants stayed and prospered, and today the town mixes its dairying and fishing with tourism, its early French history tweaked up to suit.

Lyttelton from the Port Hills (*top*). The town serves as port to Christchurch. Its Timeball Station (*above*) once signalled by the dropping of a ball a precise time by which mariners set their chronometers. The station dates from 1876, and closed in 1935, but has been fully restored since. The hills of Banks Peninsula (*below*), overlooking Akaroa Harbour.

Banks Peninsula is volcanic, and the pretty resort and fishing port of Akaroa snuggles against the wall of an ancient, sea-invaded crater. The name Akaroa means 'long harbour', and the century-old Akaroa lighthouse (*above*) which once stood sentinel at the harbour entrance, now presides over New Zealand's 'fragment of France'. The French colonised the site in 1840. Though few original traces remain, a picturesque spirit still hovers amidst its houses (*top and centre*). The wharf (*right*) is used by the Akaroa fishing fleet.

State Highway 73, west of Christchurch, leads across the plains to a ski-field at Porters Pass. At 945 metres, this is the highest road pass in New Zealand. Beyond, popular as a winter skating rink, is Lake Lyndon, then three more ski-fields. Near the next summit, the road enters the Arthur's Pass National Park. The park gives an easy entry into the wonders of the Southern Alps. It boasts sixteen peaks higher than 2,000 metres, and many walking, tramping and climbing tracks easily accessible from the road. Arthur's Pass itself, 924 metres above sea level, is named for the surveyor who found the route of the first coach road from Canterbury to Westland in the 1860s.

South from Christchurch on the inland Highway 72 is the spectacular Rakaia Gorge, and beyond it Mount Hutt, a ski-field of international class, with a long season extending from June sometimes as late as November.

South from Christchurch down State Highway 1 is the sizeable plains town of Ashburton, originally a stop-over point for coaches fording the river there. Beyond it is Timaru, the second-biggest population centre within Canterbury and a prosperous port town for both industry and agriculture.

Mount Hutt (*above*) rises abruptly. Its height and exposure to wind ensure its ski-field a five-month-long season, beginning usually in June. The Rakaia Gorge (*below*), one of South Canterbury's beauty spots, is crossed by State Highway 72 near Mount Hutt. Here the river flows swift, deep and dangerous, but transforms into the classic Canterbury Plains braided river nearer the coast.

The wheatlands and mixed-cropping of the plains gives way in South Canterbury to a pastoral economy. At Methven (*right*) a farmer loads sheep. At Temuka the cattle sale (*above and below*) attracts pedigree stock and spirited bidding.

Timaru (*left*) raised and still remembers New Zealand's greatest race horse, Phar Lap. In the late 1920s and early 1930s the huge horse beat all-comers in New Zealand, Australia and the United States, to become the richest stake winner of his era. Timaru's Atlas flour mill (*above*) and the city's historic Landing Service Building (*below*), which dates to 1867.

OTAGO

The Scots colonised the coast of Otago, founding Dunedin, the most substantial of New Zealand's nineteenth-century cities. Gold did the rest. A dry and tawny interior still preserves the relics of gold rushes amidst today's enduring occupations of sheep and cattle runs, stone and pip-fruit orchards.

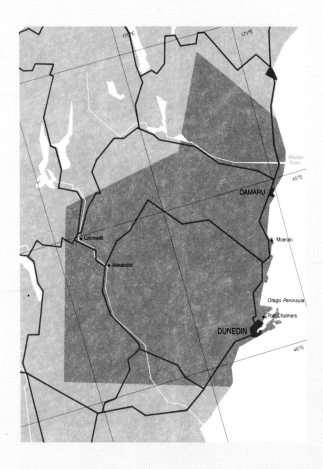

Above: *Waitaki River bridge.*
Left: *The lake at St Bathans.*

Oamaru stone at the quarry (*right*) and in neo-classical display at the Oamaru Courthouse (*below*), built in 1883. The unusually white and easily crafted limestone, is used countrywide, but nowhere to greater effect than in its home town. Optimistic nineteenth-century Scots raised in Oamaru stone such an assembly of banks, hotels, public buildings and churches that today's visitor may be struck by a discrepancy of scale: the North Otago township's modest size, the big-city magnificence of its buildings.

Vincent Pyke, an early explorer and author, described Central Otago as 'devoid of animal as of human life; where the stillness was painful in its prolonged intensity; and the only sound that greeted the ear from dawn to dusk was the melancholy wailing of the wind among the tussocks'.

That interior is first glimpsed high up the Waitaki River valley, where stands the mightiest series of hydro-electric projects New Zealand engineers have built. At Benmore, the largest of them, the penstock pipes are as big as train tunnels, and amidst arid hills a mighty earth dam holds back New Zealand's largest artificial lake.

Yet there is also a different Otago. The coastal State Highway 1 traverses rich sheep and stud farms to reach North Otago's main town of Oamaru, a substantial centre resplendent with a nineteenth-century architecture of stone.

The highway passes Moeraki Beach and the huge spherical boulders there, explained in Maori myth as spilled food kits from the sunken *Araiteuru,* an ancestral canoe. This coast was once densely settled by Maori tribes, but they had already declined at the time of European contact and a new phase, of Scottish development.

In the 1840s, company executives of the same New Zealand Company that had established settlements at Wellington, New Plymouth, Wanganui and Nelson tapped economic and religious discontent in Scotland. They floated an immigration scheme but had no site. The

Otago's coastal strip belies the provincial reputation as a tussocky desert. In a moist and misty climate, pine forest flourishes near Herbert (*above*). A rich agriculture also prevails. Feed time at 'Goodlife Farm', Herbert (*below*).

The unusually perfect Moeraki Boulders (*right*) are technically not boulders at all but 'concretions'. Their nucleus is a small fragment of shell or bone that attracts minerals like calcite, limonite or quartz, binding agents between rock grains. The 60-million-year-old 'boulders' build like pearls. The 'Organ Pipes' on Mount Cargill (*above*) result from vertical splitting in cooling basalt. The old volcano provides Dunedin's highest viewpoint over the Otago Peninsula (*below*).

Otago Peninsula, an old volcanic formation, its crater flooded to form a long and commodious harbour, had long been known to the sealers. In the 1830s, a whaling station was ensconced there. A New Zealand Company surveyor, Frederick Tuckett, saw the peninsula, liked it, and in 1844 bought most of it. Four years later the first ships of a so-called 'class colony', Scottish Free Church Presbyterians, arrived to found their city. They called it New Edinburgh, or, in a Celtic form of the name, Dunedin.

The city today is New Zealand's most distinct. It is small, supporting a population of 114,000. It is steep – seven populous hills overlooking a small amount of flat land at the base of the harbour. It has snow – the only New Zealand city to experience a regular winter snowfall. It is solid – in the nineteenth century three successive gold-mining booms, alluvial, quartz mining and finally dredging, helped establish Otago as New Zealand's richest province. The wealth did not trickle through the fingers of the Scots. They built New Zealand's most powerful commercial city, a town whose churches and university, law courts and railway station, municipal chambers and even its police station are paradigms of Victorian splendour. Residential Dunedin boasted brick terrace housing uncommon to New Zealand, and mansions of the wealthy. Most of all, Dunedin is Scottish. At the city's centre is the Octagon, a pleasant green, overlooked by a statue of Robbie Burns, Scotland's immortal bard. Nearby, in Moray Place, is the elegant Presbyterian First Church, the finest Gothic Revival church in New Zealand. The city sells kilts. It makes whisky.

A drive along the southern arm of the peninsula leads to Larnach's Castle, where the dream of recreating old Scotland assumes the dimension of a folly, and on to Taiaroa Head, the only known mainland breeding ground of the royal albatross. Public access to the colony is strictly controlled. On the peninsula's northern arm is the salty town of Port Chalmers, Dunedin's deep water harbour.

The purchaser of the Otago Peninsula, Frederick Tuckett, noted the Dunedin town site had a single fault – 'the distance from the deep water of the lower harbour'. Port Chalmers, 12 kilometres up-harbour from Dunedin, became Otago's main deep-water port. In 1882 the first shipment of frozen meat left here for Britain. Today the town is home to a fishing fleet at Careys Bay (*above*). When the container revolution of the 1960s forced every New Zealand city to find deep-water access for large ships, Port Chalmers was the obvious choice for the province's terminal (*below*).

Dunedin's nineteenth-century wealth left a legacy of substantial Victorian architecture. Otago Boys' High School (*top*), the University of Otago's clock tower (*above*), Larnach's Castle (*right*), and university buildings (*below*).

Much of Pyke's early description still fits Central Otago. In summer the region is New Zealand's hottest and driest inhabited territory, and in winter one of the coldest. Squatters ran sheep there in the 1850s. In 1861 came Gabriel Read's report of a goldfield so rich the metal could be shaken from the roots of the tussock. Thousands of fortune-seekers rushed inland. Highway 8, south of Dunedin, leads to that same destination, Lawrence, now a small farming town alongside the site of the first gold rush, Gabriels Gully.

The road leads on across the Clutha River to the town of Roxburgh, a large hydro-electric dam, and wide river flats planted out with stone fruit. Together with the irrigated valleys of nearby Alexandra, much of New Zealand's apricot crop originates here. From Alexandra, Highway 85 leads north to the historic little gold towns of St Bathans and Naseby, and to Ranfurly, service centre of the Maniototo Plain. Westward up Highway 8 lies Cromwell Gorge, site of the second great gold strike in 1862. The town of Cromwell has been relocated to the edge of a proposed lake as engineers work to complete the giant Clyde Dam hydro-electric project. Beyond that, Highway 6 shadows the old gold trail to the Arrow and Shotover Rivers, where in the 1860s Otago's fabulous series of gold strikes continued.

Sheep pens of stone (*above*) at Middlemarch, Central Otago, use the region's easily split schist. Central's haunting power arises from its absences — of everything but winter snow, summer heat, wind and sheep, seen here (*below*) amidst a fresh snowfall on the Rock and Pillar Range.

Alexandra is Central Otago's biggest town. The Clutha, in water volume New Zealand's mightiest river, was first bridged here in 1882. The stone piers of the old bridge still stand, superseded now by the arched steel crossing, built in 1958 (*top*). Near Cromwell a bungi jumper makes a death-defying leap into the Kawarau River gorge *centre* . The bungi jump was invented by New Zealand's A. J. Hackett. It needs a chasm, a carefully measured length of elastic rope, and a willing victim. Professional companies now cater for a surprisingly steady demand. Central Otago's apricot orchards are the region's chief horticultural wealth. At Roxburgh, Cromwell and at Alexandra (*right*) the trees flourish on irrigated land.

The 'Golden Progress' mine (*above left*) at Oturehua, near Alexandra, sums up Otago's nineteenth-century dreams and the twentieth-century dereliction. Gold fever began in 1861 with Gabriel Read's great find. A series of rich alluvial fields were quickly exhausted. Quartz mining began, and then steam dredges turned over river beds. Sluicing for gold from the hills of St Bathans has left amidst lunar landscapes a lovely lake (*left*). At Bendigo stand the stout stone remains of a miner's cottage (*below*). Gold is still panned at Cromwell (*above*), and elsewhere the elusive metal is again being squeezed from the land by smart technologies.

SOUTHLAND

Scottish husbandry has groomed the southern downs into New Zealand's most productive sheeplands. The main city of Invercargill prospers on the southern shore of that pastoral plenty, and beyond it, Stewart Island adds a charming full stop to the region the Maori call 'Murihiku' — 'land's end'.

Above and left: *The Catlins coast.*

One of the eleven Cathedral Caves at Waipati Beach (*right*) on the Catlins coast. The interlocking caverns are 30 metres and more high, alive with echo, and ferns that grow even from the roof. Exploration is restricted to one hour before and after low tide. The Purakaunui Falls (*below*) are another Catlins beauty spot, reached by side road off Highway 92 and a short walk. Water cascades over two rock steps before dividing into an array of cataracts.

The Otago Plateau descends to a gentler mix of rolling downs and river flats south of the Clutha River. From the coast across to the forests of Fiordland, the land is now green, the people few, and the sheep many. Some four per cent of the population produce almost twenty per cent of New Zealand's primary exports.

Balclutha, the main town of south Otago, stands on the river bank, a service town to the Clutha Valley. The river itself flows wide and swift here before forking at the alluvial island of Inchclutha and discharging to the sea. Though a little shorter than the Waikato, New Zealand's longest river, the Clutha is the country's most powerful, carrying almost twice the water volume of its North Island rival.

Highway 1 turns inland here, through rolling green downs to the prosperous town of Gore. Southland has Scottish pioneers, and they did not neglect a traditional crop. Gore is the traditional home of New Zealand's breakfast cereals, mainly oats, a fact central to the definition of eastern Southlanders as a folk raised on fresh air, porridge and fear of the Lord. Nor did they neglect a traditional drink. The strangely turreted Hokonui Hills, which stand behind Gore, are famous as the source of potent moonshine whisky, an industry that has now lapsed.

The coastal Highway 92 from Balclutha to Invercargill is a prettier route, passing through the Catlins Forest Park headquarters at Owaka and opening the park to exploration. One of the south's loveliest walks is a six-hour track beside the Catlins River. South of the river estuary the cliffs are tall white rock, facing a stormy southern ocean, and Highway 92 gives frequent access to that dramatic coast, to the lovely Purakaunui Falls, to beaches where southern cribs of stone face surf-washed beaches, and to such natural curiosities as the Cathedral Caves and a petrified forest, some 160 million years old, at Curio Bay.

Sheep near Nightcaps in the southwest of the province (*above*). Southland has the country's most extensive sheep flocks. A land once predominantly tussock-covered has been drained, enriched with lime and superphosphate, and transformed to green and highly productive sheepland. Young pony club riders at a gymkhana outside Balclutha (*below*).

The Invercargill City Art Gallery (*top*) is housed in the old Georgian-style home of philanthropist Sir Robert Anderson. The gallery holds an excellent collection of New Zealand paintings. It is sited 7 kilometres from the city centre, in the huge Anderson Park. In a city that has eschewed high-rise construction, the water tower (*right*), built in 1888, is Invercargill's tallest landmark, and also its oldest and most picturesque. The famous 'Bank Corner' (*below*) found at the intersection of Dee Street and Tay Street.

Invercargill (pop. 54,000) is the southernmost New Zealand city, its traditions Scottish and conservative, its streets broad and logically laid. The city has based a solid prosperity on the rich surrounding grasslands, and on the pioneering of licensing trusts, which channel liquor profits back to the community. One result, aided perhaps by the plentiful flat land, is Invercargill's wealth of parks, sports grounds and gardens — a ratio of reserves to citizens three times the usual urban standard. Queens Park, close to the city centre, is huge. It includes a golf course and a cricket field, lawns for croquet and bowls, gardens dedicated to the rose, rhododendron, azalea and alpine plants, playgrounds, fountains, an aviary and charming sculptures for the city's children. West of the city, on the road to the popular Oreti surf beach, are extensive sports grounds.

From Invercargill, State Highway 1 leads on to land's end at Bluff. The town huddles away from Foveaux Strait's wild weather below Bluff Hill. Its harbour is Invercargill's port, but Bluff has important industries of its own: a major freezing works, fish processing, and a specialist fleet of small boats to dredge from Foveaux Strait the delicacy for which the town is renowned — Bluff oysters. Across the harbour, at Tiwai Point, ships discharge bauxite. On that same site the constant plume from the tall Tiwai Point chimneystack marks the massive application of Lake Manapouri electricity to that imported raw material. The smelter there produces 240,000 tonnes of aluminium every year.

A navigation beacon at the entrance to Bluff Harbour (*above*). Foveaux Strait is New Zealand's roughest strip of water, and Bluff Harbour a narrow-throated haven for the ore-carriers, the cargo freighters, the oyster fleet, and the fishing boats (*below*) that use the port.

A small ferry from Bluff and air flights from Invercargill cross Foveaux Strait to Stewart Island. The visitor's first surprise is the island's sheer bulk, dominated in the north by the 980-metre Mt Anglem, but rolling away south in bush-mantled hills as far as the eye can see. It is the third-largest island in New Zealand, surpassed only by the North and South Islands.

The next surprise, given the wilderness of that interior, is the gently civilised air of Oban, the island's single village. It is set in the rock-girt but near-perfect sandy crescent of Half Moon Bay. The beauty of that beach is no oncer. The island's long coastline yields a succession of such protected sandy bays, mostly deserted.

Just 500 people live in and around Oban, and a mere 26 kilometres of road service the inhabited bays. Beyond that, only the Department of Conservation tracks penetrate an interior inhabited only by abundant bird-life, and deer.

Stewart Island was sold by the Ngai Tahu Maori in 1864. They called it Rakiura, or 'glowing sky', a name that describes both the long and brilliant sunsets of these low latitudes and the occasional shimmering bands of the aurora australis in the night sky. The island has a history of sealers, whalers and saw-millers, all come and gone. Today the local economy is made buoyant by cray and cod fishing, coupled with salmon and mussel farming in Paterson Inlet. The fishing industry imposes its own unhurried rhythm around Oban, and the many retired residents merge easily into that. There is a small tourist trade, but the island's last surprise is that the unspoiled beauty of the coast and the challenging tramping and hunting opportunities inland have remained largely undiscovered by New Zealanders or visitors to the country.

Fishing boats and pleasure craft line the jetty at Riverton (*above*). The sheltered estuary gives access to Foveaux Strait, and the harbour was once, until ship sizes increased, a competitor to Bluff. Established as a whaling station in 1835, Riverton is the oldest Pakeha settlement in Southland. Windswept Te Waewae Bay (*below*) looks across the last of Southland's green shoulders to the purple distance of Fiordland.

A Stewart Island resident walks on seaweed-strewn Halfmoon Bay after a storm (*left*). Fishing determines the island's seasonal rhythm, and the daily rhythm is geared to the tides. Boating is a way of life, and boat-sheds and ramps (*above*) line most of the bays around the island's single village, Oban. Deep and wide Paterson Inlet (*below*) provides sheltered water for a mussel-farming industry, and the first-stage launch trip for visitors who follow tramping trails across the island.

FIORDLAND

Ice 2,000 metres thick once capped Fiordland and flowed outwards to the sea. The glaciers are gone, but the ice-milled valleys remain, now partially invaded by the sea as deep fiords. Those awesome avenues, the mountains that rise rank upon rank between, and the forbidding bush give Fiordland its preternatural power. It is New Zealand's most profound wilderness.

Above: *Lake Te Anau.*
Left: *Mitre Peak, Milford Sound.*

Fiordland has been heavily glaciated at least four times during the past two million years. Glaciers a kilometre high have carved valley walls to a startlingly uniform slant and gouged the coastline into fourteen deep fiords.

In this wilderness the Maori were the first human voices, weaving stories of great beauty around its lakes and fiords and painting the symbols of magic in its caves. A battle at Te Anau drove the defeated Ngati Mamoe deep into its interiors and gave rise to 'the lost tribe' legend, one with a modern echo, for the large, flightless takahe, believed extinct, was found here in 1948.

James Cook named Doubtful and Dusky Sounds on his first circumnavigation. He entered the fiords on his second voyage of 1773, spending a month at Dusky Sound. His reports of seals drew ships from Australia, and whalers followed to complete an early history of bloody plunder.

Today this southwestern rump of the South Island is a protected wilderness. The Fiordland National Park boundaries encompass a region that is the wettest part of the country and the least populated; it has the highest waterfall and the deepest lake; there are few roads but over 500 kilometres of track. The park's 1.2 million hectares make it by far the biggest in New Zealand.

Te Anau (pop. 2,400) is the park headquarters, a tourist town where scenic flights can be arranged to the main fiords, including those, like Dusky Sound, where access is otherwise difficult. The town lies on the eastern margin of Lake Te Anau, the South Island's largest lake. Like the region's other big lakes, it was originally a glacier bed dammed by moraine debris. A wildlife reserve near the town gives the chance to meet the takahe, and a launch crosses the lake to the Te Anau Caves.

A waterfall in Doubtful Sound (*opposite*). In 1770, on his first voyage to New Zealand, an anxious James Cook decided not to enter this crack in the western seaboard, and named it accordingly. On that same voyage, Cook came upon Dusky Sound (*above*) as evening drew in, and the cautious Yorkshireman stood away from this fiord also, naming it 'Duskey Bay'. The fiords, like Lake Te Anau (*below*), are channels carved by ice.

FLORA AND FAUNA

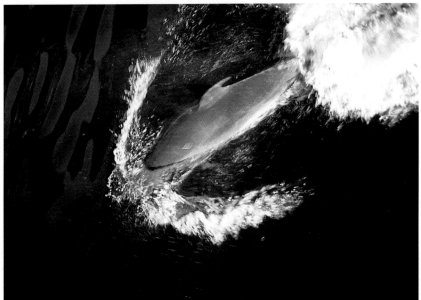

In Fiordland wilderness, nature's vignettes abound. A harsh climate shrinks the size of the region's flowers but not their beauty (*above*) nor that of an exquisite native orchid (*top*). A Buller's mollymawk (*top left*) eyes human intruders in Dusky Sound, and a dolphin sports on the pressure wave in front of a sightseeing boat at Milford Sound (*left*). In one of the wettest climates on earth, fungal fruitbodies abound. The 'golfball' fungus (*below left*) and the delicate crimson agarics (*bottom left*) are both found in Fiordland beech forest. A seal in Dusky Sound (*below*) remains happily blank to a history of carnage, with clubs, by last century's sealers.

From the town a road skirts the lake and heads to Milford Sound. It crosses the Divide and passes the entrance to the three-day Routeburn Track to Lake Wakatipu. Past the Mirror Lakes, the grasslands and beech forest of the Eglinton Valley, the road has already been a striking route. Now it drops into the Hollyford Valley, and becomes awesome. The sheer U-shaped walls of an old glacier route rise almost sheer on either side, shrinking the human observer to a dot. In wet weather these same walls are laced with myriad waterfalls dropping hundreds of metres to the valley floor. A turn-off leads to Hollyford Camp, and beyond it the Hollyford Track opens up a four-day tramp down Fiordland's longest valley.

The main road continues on to the Homer Tunnel, boring 1,240 metres through the base of a mountain barrier to emerge in the Cleddau Valley and on down switchback curves to Milford Sound.

The famous Mitre Peak dominates the Sound. Rising 1,692 metres straight out of the water, the peak and its still, reflective setting converge in natural perfection. Unsurprisingly, it is the most photographed pinnacle in New Zealand. The Bowen Falls, thundering out of a hanging valley near the base of the fiord, are easily reached on foot from the Milford Hotel, but further exploration is best by boat, through the mist of plunging waterfalls, past ice-striated rock walls to within sight of the ocean.

A second route into the Sound is the 54-kilometre Milford Track, New Zealand's most famous walk. Guided groups or the independent 'freedom walkers' usually do the same

The Bowen Falls at Milford Sound (*above*) and a view of Mitre Peak (*below*). The Sound is Fiordland's most beautiful, and its most accessible.

In the Mararoa Valley (*top*), between Lake Te Anau and Lake Wakatipu, lie the Mavora Lakes, accessible to trampers by a narrow farm road. New Zealand's most famous walk, the Milford Track, reaches its highest point at Mackinnon Pass (*above*). Beyond, in early morning mist, is Mount Elliot and the short but thick ice-flow of the Jervois Glacier. A three-wire river crossing on the Dusky Sound track (*right*). The tramp to Dusky Sound is one of Fiordland's hardest. On a seven-day journey, trampers carry all their supplies, in and out, across rivers prone to flooding.

tramp in four days, beginning at Glade House on the northern tip of Lake Te Anau. The track follows a forest-shaded river beneath the towering walls of Clinton Canyon, then rises to alpine country and the saddle named for Quintin McKinnon, who pioneered the route in 1888. Beyond Mackinnon Pass, dropping 580 metres in three stupendous leaps down a sheer rock wall, are the Sutherland Falls, the highest waterfall in New Zealand and the fifth highest in the world. The track goes on down the gigantic-sided Arthur Valley to the aptly named Sandfly Point. From there, a launch takes trampers up the Sound to finish at the Milford Hotel.

Twenty kilometres by road from Te Anau is Lake Manapouri, the deepest lake in New Zealand and, with its many islands and the forests descending to its waterline of beaches and coves, perhaps the most beautiful.

Originally the lake's outlet was the Waiau River to the south, but its West Arm is poised above Doubtful Sound, and engineers planned a hydro-electricity site there. The project went ahead in the 1960s with the construction of a powerhouse and penstock pipes, all underground, and a water-race spilling into the Sound. The lake was to be raised 25 metres, but in the first outing of New Zealand's strong conservation lobby, this was amended to eight metres, and then to no rise at all. A launch takes tour parties to West Arm, and a bus spirals down to the power station. A bus tour also leaves from here to visit Doubtful Sound far below.

The road runs south to Lake Monowai, whose shores, littered still with stumps after the level was raised in 1925 for a power project, served as the warning for Manapouri. The timber and farming town of Tuatapere is nearby, and from there a road leads to the last of Fiordland's easily accessible lakes, Lake Hauroko, which still preserves on an offshore island the centuries-old sitting burial site of a high-ranking Maori woman.

The Manapouri power station is one of New Zealand's engineering masterpieces. Every minute thousands of tonnes of Lake Manapouri water drops vertically down penstocks to drive generators housed in the rock over 200 metres below. The water is discharged along a 10-kilometre tailrace tunnel to spill into Doubtful Sound. The station is open to visitors, and a bus leaves West Arm to drive down the corkscrew access tunnel to the powerhouse (*below*). In that clinically humming place, seven generators are producing up to 700 megawatts of power, making the station New Zealand's second largest.

MOUNT COOK AND THE LAKES

New Zealand's greatest peaks and prettiest lakes are arrayed in the South Island's most scintillating landscape. To the north, New Zealand's highest mountain, Mount Cook, chisels the sky. To the south, two resort towns cater for more lake-level taste: Wanaka quiet and contemplative, Queenstown effervescent with jetboat, cableway and ski-field thrills.

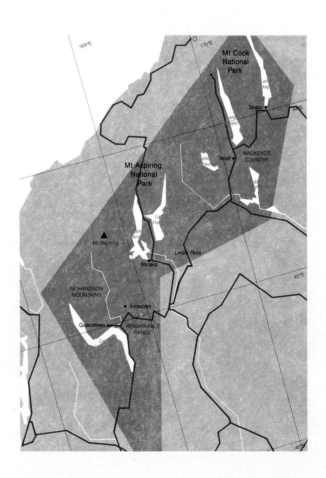

Above: *Lake Wanaka.*
Left: *Queenstown moorings.*

145

At Tekapo stands the bronze memorial (*right*) to the sheepdogs of the Mackenzie Country, 'without the help of which the grazing of the mountainous country would be impossible'. The tribute is therefore cast wider than the popular story that the sculpture honours the sheep-stealer James McKenzie's dog Friday. The Church of the Good Shepherd (*below*), built of local stone, stands on the edge of Lake Tekapo as a memorial to Mackenzie Country pioneers.

The northernmost entrance to the South Island's sublime heart lies through the Mackenzie Country. It is named, though misspelled, for James McKenzie, a Gaelic-speaking shepherd caught driving a thousand stolen sheep onto the highland in 1855. McKenzie and his dog Friday became folk heroes. The Mackenzie Country, a high mountain-bounded plain unknown to Europeans at the time of the shepherd's capture, added a lost-world dimension to the myth.

Silence, space and the shimmer of snowy mountains still invest the Mackenzie Country with that lost-world air. Nor is the spell broken by Lake Tekapo, its turquoise water made milky by the glacier-ground rock flour of the mountains behind. Beside the lake stands the stone Church of the Good Shepherd, its altar window admitting a famous view of the alps.

The road crosses a low dam, part of a hydro-electricity project that knits all the Mackenzie Country lakes into a canal-connected unit to generate power, provide irriagtion water and control waterflow for the large Waitaki River power stations.

As Highway 8 continues on to Lake Pukaki, the vista of Mount Cook opens for the first time. Samuel Butler, the English scientist, novelist and essayist who founded a high-country sheep run in New Zealand, once wrote: 'If a person *thinks* he has seen Mount Cook, you may be quite sure he has not seen it . . . There is no possibility of a mistake.' The mountain is 3,764 metres high, New Zealand's king peak, but it is not alone. To its right is Mount Tasman (3,498 metres), and around it some 150 summits topping 2,000 metres. At Lake Pukaki, a road turns off to Mount Cook village, the Hermitage Hotel and the headquarters of Mount Cook National Park.

The Mount Cook lily (*above*) blooms in spring and summer. It is not a true lily, but a giant member of the Ranunculaceae, or Buttercup family. Mount Cook (*below*), New Zealand's highest peak, seen from the Hooker Valley.

The terminal lake and icebergs of the Tasman Glacier (*above*). Though now diminished from the monster glacier that pushed deep into the Mackenzie Country, the Tasman ice flow is still formidable, 3 kilometres wide in places, and 29 kilometres long. It is New Zealand's largest glacier, and the largest of the world's temperate-region glaciers. The four-hour walk up the Hooker Valley (*right*) is one of the Mount Cook National Park's most popular trails. The walk crosses the Hooker River twice to reach the terminal face of the Hooker Glacier.

The park is the ice- and snow-climbing mecca of Australasia, but there are numerous walking tracks, an ice traverse over the Copland Pass into Westland, and a bus trip and short walk to overlook the 29-kilometre Tasman Glacier, New Zealand's longest. A ski-plane takes sightseers and skiers to the head of the mighty ice flow.

Beyond the Mount Cook turn-off, Highway 8 passes through Twizel, built by the government as a construction town for the Upper Waitaki hydro-electric projects, and the settlement of Omarama. In 1848 the South Island's largest tribe, Ngai Tahu, parted with eight million hectares of their land for just £2,000. Promises of large reserves and preservation of traditional food-gathering places did not eventuate, and Ngai Tahu bitterness found expression under the leadership of a prophet, Te Maiharoa. At Omarama he founded a settlement on runholders' land. After an armed confrontation the village was dispersed in 1879.

Lake Hawea, blemished by a scheme to raise the level 20 metres and ensure the Clutha River power projects an adequate winter water supply, is still a good fishing lake. The adjacent Lake Wanaka remains undisturbed. Island-dotted, in a setting planed smooth by an ancient glacier, it is a beautiful lake. The pretty township of Wanaka gazes across pellucid waters to rugged mountains. It has developed into a resort town with summer water-sports, winter skiing at Treble Cone and skating on Lake Diamond. The township is also headquarters to the Mount Aspiring National Park, which lies directly west. Though tracked, the park's long and forbidding mountain valleys and rugged ice-gouged interior make it a place for experienced trampers. The centre of the park is Mount Aspiring (3,035 metres), its distinctive spire earning it the title of New Zealand's Matterhorn.

Rising beyond the Ahuriri River, are the Omarama Spires (*above*), a cliff-face weathered into dramatic columns. Lake Ohau (*below*) is the most southern of the Mackenzie Country's snow-fed lakes.

Highway 8 leaves the Mackenzie
Country at Lindis Pass (*above*) for
Otago's lake district. The road rises
to 970 metres above sea level,
following an ancient route of Maori
fishing and fowling expeditions.
Lake Hawea (*right*) is the least
populated of the three Otago tourist
lakes, its beaches and shoreline
sacrificed to a hydro-electricity
project that raised the level in 1958.
Yet it is favoured by many as an
excellent fishing spot for trout and
salmon.

Highway 6 to Queenstown takes up the old gold trail of the 1860s as miners prospected on from the Clutha River strikes up the Kawarau, Arrow and Shotover Rivers. Arrowtown arose as an instant metropolis after a rich gold strike on the Arrow River in 1862. It is now a tourist centre with picture-postcard poplars, refurbished miners' cottages, nineteenth-century shop façades and a good local museum. Two months after that strike, two shearers discovered gold in the Shotover River. Miners won 72,000 ounces of gold there within the first three months and dubbed it 'the richest river in the world', but it is harsh, gorge-ridden territory, opened now to view by jetboats up the river.

Queenstown, originally a gold settlement, is now the premier South Island resort town, and the most cosmopolitan. The famed powder snows of its Coronet Peak ski-field draw skiers from around the world. The town is set on the shore of Lake Wakatipu, second largest of the South Island lakes and in legend the most dramatic. Maori tradition tells of a sleeping giant immolated here, his knees drawn up in agony, his heart continuing to beat even as the fire burned deep into the earth and the depression filled with water. That heart beats still. The lake rises and falls as much as six centimetres every five minutes, an effect known to science as a seiche and explained by fluctuating air pressure.

The lake setting is spectacular, from water level, aboard the old steamer *Earnslaw*, or from Bobs Peak, a 445-metre summit overhanging the town and served by a gondola cableway. The Remarkable Range rises rugged from the eastern shore, the Humboldt Mountains sheer from the western edge, a panorama entirely fit for a queen's town.

The beautiful Lake Wanaka is edged by quiet coves like Glendhu Bay (*above*) and overhung by mountains. The view from the road to Treble Cone ski-field (*below*), looking east.

Heli-skiing opens up to skiers the snows of the Richardson Mountains, north of Queenstown. A helicopter lands them near the summits (*above*) and they ski the virgin piste, (*right*). The Matukituki Valley (*below*) is a broad pathway into the Mount Aspiring National Park.

At Skippers Canyon (*left*) a promontory of stone now silently guards a valley that once rang with prospectors' shouts. The last of the fabulous alluvial gold strikes that enriched Otago in the 1860s ended with rushes to this bleak canyon and the Arrow River. The once-seething gold settlement of Arrowtown (*above*) has been preserved and reconstructed as a period piece. The town is a popular tourist destination.

Whitewater rafting on the Shotover River (*above*). The powerful shallow-draught jetboat was the invention of New Zealand's Bill (later Sir William) Hamilton. Jetboats based at Queenstown take tourists on exciting rides through the gorges of the Shotover River (*right*). High above Lake Wakatipu, the lookout from the Remarkables ski-field (*below*) opens up views of the mountain vastness around Queenstown.

154

Visitors await Queenstown's 'Lady of the Lake' the 1912 steamer TSS *Earnslaw*, (*left*) which plies Lake Wakatipu. One of New Zealand's most distinctive panoramas is from Bob's Peak at the head of the gondola cableway, looking across Queenstown and Lake Wakatipu toward the Remarkables Range (*below*). A shepherd demonstrates handling skills at the Mount Nicholas sheep station on the southern side of Lake Wakatipu (*above*).

WEST COAST

Gold and coal have fueled the boom and bust history of Westland. Irish blarney has stoked its tall tales. On this lean coastal strip, lodged between the salt haze of its breaker line and the wall of the Alps, folklore and physical magnificence combine. 'The Coast', as it is commonly called, is a place apart, 'the Coasters' a breed of their own.

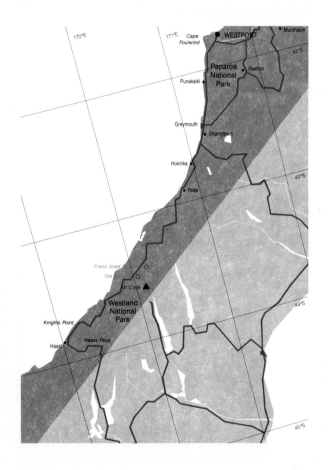

Above: *The Southern Alps.*
Left: *Fishing boats at Westport.*

The Haast Pass road, the lowest but owing to engineering difficulties the last completed of the Alpine crossings, has linked Otago to Westland since 1960. Winter ices the banks of the Makarora River, just east of the Pass's 564-metre summit (*top*). Bulls near Haast (*right*) brave the same wintery conditions. A tourist bus (*above*) lets its passengers taste the delights of a recent snowfall.

158

From Jackson Bay in the south to Karamea in the north, a simple geography prevails. For 500 kilometres the surf beats incessantly on a scabrous coastline. Parallel to that shore run the Southern Alps. Between the two, seldom more than 40 kilometres wide, are the coastal flats and densely bushed foothills of Westland.

The Haast Pass is its southern entrance. The landscape undergoes dramatic change from the tawny grassland of Otago to the dense forest, the rushing creeks and moody rivers of a land drenched by heavy rainfall. The pass peaks at 564 metres and descends to the tiny settlement of Haast, crossing the Haast River on the longest single-lane bridge in the country. Ahead lie the two mighty glaciers of the Westland National Park. They nose in ponderous splendour down from alpine neves, through shattered valleys to a bushline just 300 metres above sea level. In the world's temperate regions, few glaciers reach so low and are so accessible.

A short track leads to the Fox Glacier's terminal face. Longer walks ascend to lookout points that take in more of the glacier's 13-kilometre length.

West of Fox township is Lake Matheson, smooth mirror to the alpine spectacle behind. The national park includes New Zealand's mightiest peaks, Mounts Cook and Tasman, and the western road to Gillespies Beach provides excellent views.

Highway 6 twists through glacial moraines to the 11-kilometre Franz Josef Glacier, Lake Mapourika and the turn-off to Okarito. The lagoon there is nesting place to the kotuku, the white heron, but special permission is required to visit.

Stories and stunning landscapes combine in Westland. Knights Point, (*below*) was named by a surveyor of the last, Haast-Paringa section of Highway 1's round-the-island route, opened in 1965. Knight was the surveyor's dog.

The Fox River headwaters issue from an ice cave on the face of the Fox Glacier (*right*). The water tunnels through the ice from a spring further up the valley. The 13-kilometre glacier (*below*) descends from vast snowfields on the western flanks of New Zealand's greatest peaks. Though shrinking imperceptibly overall – the estimated rate is 3 kilometres every century – the Fox's daily flow is an impressive 1.5 metres.

Nature walks abound in the Fox area. The region's torrential 4,800mm rainfall is evidenced by brimming streams and mossy forest (*left*). Lake Matheson (*above*), originally a sink hole left by a vast piece of melting glacier ice, is famous for its mirror image reflections of Mounts Tasman and Cook.

Westland's plains are narrow and frequently waterlogged, but with the ending of the gold boom and the diminution of coal mining, farming has become the major coastal industry. Farmland at Fox (*right*). Beyond Franz Josef the land shows the distinctive Westland mix of plain, bushed hills and alpine backdrop (*below*).

Hokitika was once Westland's main gold-rush town, and something of that atmosphere lingers still, though the perilous port that once admitted fortune-seekers by the tens of thousand is now closed. Gold fever populated the Coast from 1864. Rich alluvial fields were first discovered in that year and for the next quarter-century, sluicing, quartz stamping and river dredging sustained the Coast as New Zealand's richest gold producer. Nearby is the pretty Lake Kanieri, and inland from there are rich dairying and stud-farming plains.

Highway 6 crosses the Arahura River, the major source of transluscent nephrite, the jade known locally as greenstone. It was prized by the Maori for ornaments and weapons of war, trading expeditions and occasionally war parties coming from the North Island to seek it from the local Ngai Tahu Maori.

Paroa marks the turn-off to Shantytown, a tourist reconstruction of a gold-rush town. Wildlife is on display at a nearby park, and further inland is Lake Brunner, named for Thomas Brunner, who became, in a nightmare 550-day journey of 1846–48, the first European to explore the length of Westland.

Greymouth (pop. 8,000) is Westland's main town, and a river port. Gold sparked it, coal sustained it, and timber and agriculture industries make it still a busy centre. Highway 7, up the Grey River valley, passes a succession of once-strong coal towns. At Blackball in 1908 the miners struck for a half-hour lunch break. Their win led to formation of the socialist Miners' Federation, a driving force in the emergence of the New Zealand Labour Party. Workingmen's clubs in the valley still display photographs of the first Labour Prime Minister, Michael Joseph Savage.

Hokitika's statue of Richard John Seddon (*above*) recalls the coast's days of power and glory. 'King Dick' was a Coaster, and Liberal Premier of New Zealand 1893 to 1906. The Grey River (*below*) at Greymouth.

HISTORIC REMAINS

First as top gold producer, then as the prime source of coal, the Coast was a past engine of New Zealand's growth. By the turn of the century most of the gold was gone. After World War II New Zealand turned to cheap imported oil for its energy, and the Coast was left with its relics. The Brunner mine (*above*) was once one of the most productive in New Zealand. Gold mining displays at Ross (*right*) recall the town's 1860's pre-eminence as one of the country's richest alluvial fields. At the reconstructed settlement of Shantytown a Kaitangata-class steam train (*below*) hauls visitors a short distance to old Chinese diggings.

The highway leads on past the ghost town of Waiuta to Reefton, once called Quartzopolis for the quantity of gold-bearing quartz crushed here. It was important enough in its day to become, in 1888, the first town in New Zealand lit by electricity.

Coastal Highway 6 continues to Punakaiki and the extraordinary Pancake Rocks, tall and layered limestone stacks with surge pools slapping menacingly and blowholes detonating in heavy weather. The roadside visitor centre has an excellent audio-visual presentation describing the rocks' formation. The area inland is part of the Paparoa National Park, with tracks into spectacular limestone country.

Westport stands at the head of the Brunner River, the main town of north Westland. It is the Coast's biggest port, and its wharves, coal mining and a large cement works give it a bustling frontier air. Beyond the town, high on an escarpment of the Paparoa Range, are the settlements, some now derelict, that once fed high-quality bitumous coal to New Zealand. Highway 67 leads on to Karamea, a dairying centre and the Coast's final town, but not the last of its hauntingly beautiful places. North, amidst primal forest, is the Oparara Arch, a great river-carved cavern, and at the lovely Kohaihai River mouth the road ends at the West Coast terminus of the Heaphy Track.

The main highway turns inland from Westport to follow the deep cleft of the Buller Gorge past Inangahua to Murchison. Both towns have been victims of large earthquakes. In 1968, Inangahua was close to the epicentre of a magnitude 7 quake. In 1929 an earthquake measuring 7.9 on the Richter scale destroyed much of Murchison. It has since been rebuilt as a centre to the surrounding farming district.

Reefton's old and new public buildings (*above*) and war memorial (*below*) reflect the town's substantial past and still-thriving present.

165

Westland's moody coastline near Punakaiki (*right*) and the spectacular Pancake Rocks (*below*). In primeval seas, carpets of calciferous debris from tiny sea creatures were regularly overlaid, amidst global tempests, by a drizzle of sand and sediment. That seabed of alternating limestone and sandstone, raised finally to the surface by tectonic force, has weathered differentially, creating the unusual layering.

On a coastline lacking natural harbours, Westport, at the mouth of the Buller River, is Westland's major river port. Fishing boats (*left*) find haven here, and coal wagons attest to a still-active export trade of high-quality bitumous coal (*below*). The Gates of Remembrance (*above*) stand at the entrance to the town's central domain.

NELSON

Nature has reached into the most benign of bags to bestow charms upon the Nelson region. The beaches are golden, the valleys green, the lakes slender and the mountains ancient, a cross-section of New Zealand's most alluring features all condensed into this single corner of the South Island. Nelson, the city in its midst, is appropriately small, pretty and self-contained, hemmed about by mountain ranges and deep bays.

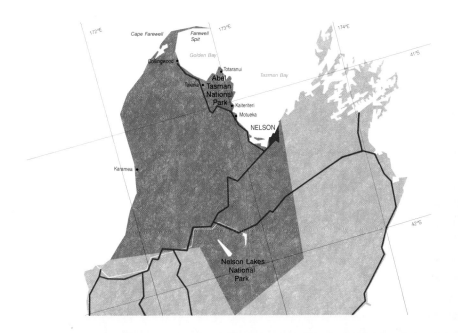

Above: *Nelson foreshore.*
Left: *Collingwood.*

Old cottages at Nelson (*above*) and the still older Broadgreen (*below*), which dates to 1855. The cobb house uses an earth construction common to New Zealand's pioneers, and its interior features period furniture. If mellow age is characteristic of Nelson city, so too are its close ties to the sea and the beaches of Tasman Bay (*right*).

After a hard start in 1842 as the New Zealand Company's first settlement in the South Island, Nelson (pop. 45,000) is now close to civic perfection. Elegant, modern high-rise exists, but in essence it's a two-storey town of Victorian shops rising gently to the Anglican cathedral upon the hill, of trees and gardens, and fishing boats bobbing in an intimate harbour.

The popular reference to 'sunny Nelson' is soundly based on statistics that place it near the top of the national sunshine tables. The climate is Mediterranean. Unsurprisingly therefore, the valleys beyond Nelson support the second-biggest concentration of apple orchards in the country, the majority of the crop destined for export. The same valley culture seems particularly suited to craftspeople and artists. In galleries within Nelson and in the townships beyond, their work is on display.

Northwest of Nelson, along Highway 60, Upper Moutere produces excellent chardonnay, riesling and sauvignon blanc wines. Further along the highway, at Motueka, hops are the traditional crop.

The cultivated lowlands are, however, only part of Nelson's attraction. No other part of New Zealand has so many forest parks, national parks or nature reserves.

The Nelson Lakes National Park lies southwest of Nelson, centred about the beautiful Rotoroa and Rotoiti lakes. Fishing, boating and hunting excursions can all be arranged from St Arnaud township, and the park offers tramping and climbing routes that give a great sense of height without much technical difficulty. Two ski-fields operate in winter.

Motueka's sunny flats and river valleys support every kind of agriculture from sheep farming (*above*) to tobacco (*below*). Along with hops, tobacco is a traditional crop, dating back to 1888. The region became New Zealand's main tobacco producer, with processing factories in the township. When government taxes first fell heavily on the crop fifteen years ago, growers were paid to reduce production.

At Nelson wharves, woodchips from New Zealand beech forest await shipment to Japan (*above*). In the Korere Valley, south of Nelson, exotic forest flourishes (*right*). Lake Rotoroa (*below*) is the larger of the twin slim lakes of the Nelson Lakes National Park. The park offers fishing, boating, pretty waterside walks, and tramping trails through beech forest.

Harwoods Hole is the most sublime single feature of the Abel Tasman National Park. At the summit of the Takaka Hill, a side road then a bush track lead on to the largest vertical sink hole in the Southern Hemisphere. At its mouth the shaft is wide enough to swallow a ship. The park encompasses most of the headland separating Tasman and Golden Bays, and is best known for the succession of golden beaches scalloping its shoreline. The camping ground at Totaranui is a base for the coastal walks. The park commemorates the Dutch explorer Abel Tasman, who, on 13 December 1642, was the first European to sight Aotearoa. Six days later, his two vessels *Heemskerck* and *Zeehaen* anchored in these sheltered waters. A longboat plying between the two ships was attacked by Maori, who killed four sailors. Tasman named his first anchorage Murderers Bay and set sail up the western coast never to set foot on the country. His unhappy name for the area was later changed to Golden Bay.

Roads lead out from Motueka to the North-West Nelson Forest Park. The region has some of New Zealand's oldest rock, ice-milled to an undulating tussock-clad tableland, but with occasional peaks. The mountain flanks support primitive plant species like the grass-tree *Dracophyllum*, and are riddled with caves. Rock-hounds take their pick from many-coloured silica, serpentine, scheelite, dolomite, marble and precious metals.

Another park access lies beyond the Takaka Hill. A steep switchback road ends at the artificially impounded Cobb Lake, its waters plummeting down 2,000 metres of pipe to the penstocks of the Cobb power station below. From the lake excellent walks lead to Peel Ridge and to trilobite territory − half-billion-year-old fossils embedded in rock.

The golden sands and wide sweep of Kaiteriteri Beach (*above*) make it one of Nelson's most popular holiday spots. Fluted limestone is a feature of the Takaka Hill north of Motueka (*below*). The immense hill is known locally as 'Marble Mountain' in reference to veins and outcrops where the limestone has hardened into marble.

The scalloped coves of the Abel Tasman National Park (*right*) give the smallest of the country's national parks an easy claim to being the prettiest. At Awaroa Bay (*below*), south of Totaranui, sea-sculpted rocks add to the pleasure of the coastal walk.

Takaka is the main market centre to the dairylands of Golden Bay. Beyond it, on the banks of the Aorere River, Collingwood grew as the boom town of New Zealand's first gold rush in 1856. Further up the Aorere Valley is the start of the Heaphy Track, a 77-kilometre, four-day trek up through open beech forest, across red tussock and down through subtropical bush to the West Coast. It is named for Charles Heaphy, a New Zealand Company draughtsman and artist who set out from Nelson in 1846 to make the first European exploration of the South Island's west coast.

From Collingwood, strangely shaped limestone hills point the way to Puponga, once a coal-mining town and now semi-deserted. Beyond that the South Island ends in New Zealand's loneliest strip of sand, Farewell Spit.

The spit is a thin crescent of sand curving 22 kilometres out to sea, its high dunes pummelled by surf on the ocean side and tapering away on the inner shore to a vast wetland of intertidal flats. New Zealand's biggest population of migratory wading birds – mainly knots, godwits and turnstones – gathers here between September and April. Black swans are another abundant species.

A four-wheel-drive bus departs regularly from Collingwood for the lighthouse at the spit's end. Interested bird-watchers who prefer to walk can trek four kilometres up the Golden Bay side, and back down the Ocean Beach. Beyond that, access is restricted. The spit is a highly protected nature reserve, one of New Zealand's most important wetlands.

Farmland near Cape Farewell (*above*) and the inner shoreline of Farewell Spit (*below*).

Overleaf: Farewell Spit, stretching 22 kilometres out to sea.